Superpave Mix Design

Asphalt Institute
Superpave Series No. 2 (SP-2)
Third Edition
2001 Printing

The Asphalt Institute can accept no responsibility for the inappropriate use of this manual. Engineering judgment and experience must be used to properly utilize the principles and guidelines contained in this manual, taking into account available equipment, local materials and conditions.

All reasonable care has been taken in the preparation of this manual; however, the Asphalt Institute can accept no responsibility for the consequences of any inaccuracies which it may contain.

Printed in the U.S.A.

The Strategic Highway Research Program (SHRP) was established by Congress in 1987 as a five year, $150 million research program to improve the performance and durability of United States roads and to make those roads safer for both motorists and highway workers. $50 million of the SHRP research funds were used for the development of performance based asphalt specifications to directly relate laboratory analysis with field performance.

Superpave™ (*Superior Performing Asphalt Pavements*) is a product of the SHRP asphalt research. The Superpave system incorporates performance-based, asphalt materials characterization with the design environmental conditions to improve performance by controlling rutting, low temperature cracking and fatigue cracking. The three major components of Superpave are the asphalt binder specification, mixture design and analysis system, and a computer software system. To date, the asphalt binder specification and the mixture design system are working satisfactorily. The mixture analyses procedures, performance models, and computer software systems are still being developed.

This manual gives detailed consideration to the Superpave mix design procedures. Full details of the Superpave asphalt binder specification and tests are found in the Asphalt Institute publication *Performance Graded Asphalt Binder Specification and Testing,* Superpave Series No. 1 (SP-1).

The Superpave mix design system is being further developed and evaluated for incorporation into federal, state, and local specifications. This manual contains several revisions to the first and second editions. Some of the more significant revisions include changes to:

▲ Mixture aging procedures
▲ Design number of gyrations (compactive effort)–N_{design} table
▲ Volumetric property specification
▲ Intermediate and complete analysis, Level 2 and Level 3 respectively, are no longer discussed.

This manual is current at the time of printing and additional changes will be incorporated into new editions as they are required. Readers are encouraged to keep informed of Superpave technology changes and specification implementation activities through industry and agency forums. Before designing a mixture, the designer should familiarize himself with any agency-specific requirements.

Much of the material contained in this manual was produced by the Asphalt Institute under contract with the Federal Highway Administration (FHWA) as the National Asphalt Training Center (NATC) for Superpave technology. The NATC developed and conducted weeklong training courses in Superpave binder and mixture technology, and was administered by the FHWA's Office of Technology Applications. Superpave™ is a trademark of the Strategic Highway Research Program.

CONTENTS

List of Figures

List of Tables

CHAPTER 1

BACKGROUND

Introduction Asphalt mixtures have typically been designed with empirical laboratory design procedures, meaning that field experience is required to determine if the laboratory analysis correlates with pavement performance. However, even with proper adherence to these procedures and the development of mix design criteria, good performance could not be assured. Also, specifications proliferated as agencies determined that they could make minor changes to established criteria and gain some assurance of performance.

In 1987, the Strategic Highway Research Program (SHRP) began developing a new system for specifying asphalt materials. The final product of the SHRP asphalt research program is a new system called Superpave, short for *Superior Performing Asphalt Pave*ments. Superpave includes a new system for selecting and specifying asphalt binders and has detailed mineral aggregate requirements. A totally new hot mix asphalt (HMA) mix design procedure is included.

The Superpave system, as originally envisioned by the SHRP research team, was intended to use increasingly complex tests and specifications as traffic levels increased. For pavements with moderate traffic, volumetric mix design, with corresponding component material requirements, was determined to be the most appropriate, cost-effective system to ensure reasonable pavement performance. For pavements with higher traffic loading, such as urban interstate highways, the researchers determined that an additional margin of safety was required to ensure pavement performance. For these high traffic volume projects, a series of mixture tests would be performed to generate measurements of fundamental mechanical properties of a compacted mixture specimen. The original intent was to input these mixture properties into performance models developed during SHRP. The anticipated output was a prediction of various forms of pavement distress as a function of time or traffic. However, research under the direction of the University of Maryland determined that some of the performance tests and prediction models were flawed. Work is underway to correct these problems. At some future time, computer software that integrates all the system components and produces performance predictions may be developed and incorporated as a standard method in Superpave mix design.

The Superpave binder specification and mix design system includes various test equipment, test methods, and criteria. The unique feature of the Superpave system is that it is a performance-based specification system. The tests and analyses have direct relationships to field performance. The Superpave asphalt binder tests measure physical properties that can be directly related to field performance through engineering principles. The Superpave binder tests are also conducted at temperatures that are encountered by in-service pavements.

The objective of this manual is to describe the asphalt mix design portion of the Superpave system. A Superpave mix design involves selecting asphalt and aggregate materials that meet the Superpave specifications, and then conducting a volumetric analysis of HMA specimens compacted with the Superpave gyratory compactor. Also provided in the manual is background information on the behavior of asphalt mixtures and materials, and a comparative discussion of traditional and Superpave asphalt binder specifications and mixture design systems. See *Performance Graded Asphalt Binder Specification and Testing*, Superpave Series No. 1 (SP-1) for full details of the Superpave asphalt binder specification and tests.

Earlier editions of this manual included a brief description of the mixture analysis procedures. That information has been deleted from this publication; an in-depth presentation of the performance tests and analysis procedures may be part of a future publication.

Asphalt Mixtures

Asphalt concrete (sometimes referred to as "hot mix asphalt" or simply "HMA") is a paving material that consists of asphalt binder and mineral aggregate. The asphalt binder, either asphalt cement or modified asphalt cement, acts as a binding agent to glue aggregate particles into a dense mass and to waterproof the mixture. When bound together, the mineral aggregate acts as a stone framework to impart strength and toughness to the system. The performance of the mixture is affected both by the properties of the individual components and their combined reaction in the system.

➤➤ Asphalt Binder Behavior

Three asphalt binder characteristics are important in asphalt mixture performance: temperature susceptibility, viscoelasticity and aging. Asphalt's properties are *temperature susceptible* – asphalt is stiffer at colder temperatures. That is why a specified test temperature must accompany almost every asphalt cement and mixture test. Without specifying a test temperature, the test results cannot be effectively interpreted. For the same reason, asphalt cement behavior is also dependent on time of loading – asphalt is stiffer under a shorter loading time. The dependence of asphalt cement behavior on temperature and load duration means that these two factors can be used interchangeably. That is, a slow loading rate can be simulated by high temperatures and a fast loading rate can be simulated by low temperatures.

Asphalt cement is a *viscoelastic* material because it simultaneously displays both viscous and elastic characteristics. At high temperatures (e.g., > 100°C), asphalt cement acts almost entirely as a viscous fluid, displaying the consistency of a lubricant such as motor oil. At very low temperatures (e.g., < 0°C), asphalt cement behaves mostly like an elastic solid, rebounding to its original shape when loaded and unloaded. At the intermediate temperatures found in most pavement systems, asphalt cement has characteristics of both a viscous fluid and an elastic solid.

Asphalt is chemically organic and reacts with oxygen from the environment. *Oxidation* changes the structure and composition of the asphalt molecules. Oxidation causes the asphalt to become more brittle, leading to the term oxidative or age hardening. Oxidation occurs more rapidly at higher temperatures. A considerable amount of hardening occurs during HMA production, when the asphalt cement is heated to facilitate mixing and compaction. That is also why oxidation is more of a concern when the asphalt cement is used in a hot, desert climate.

The characteristics of asphalt cement under varying temperatures, rates of loading, and stages of aging determine its ability to perform as a binder in the pavement system. The tests and specifications used to measure and control these characteristics in the Superpave system are discussed in *Performance Graded Asphalt Binder Specification and Testing*, Superpave Series No. 1 (SP-1).

➤ ➤ Mineral Aggregate Behavior

A wide variety of mineral aggregates are used to produce HMA. *Natural* aggregates are simply mined from river or glacial deposits and are used without further processing to manufacture HMA. These are often called *bank-run* or *pit-run* materials. *Processed* aggregate has been quarried, crushed, separated into distinct size fractions, washed, or otherwise processed to achieve certain performance characteristics of the finished HMA. *Synthetic* aggregate is any material that is not mined or quarried and is often an industrial by-product, such as blast furnace slag. Occasionally, a synthetic aggregate will be included to enhance a particular performance characteristic of the HMA. For example, lightweight expanded clay or shale is sometimes used as a component to improve the skid resistance properties of HMA. Also, an existing pavement can be removed and reprocessed to produce new HMA. Therefore, *reclaimed asphalt pavement* or *RAP* is an important source of aggregate for asphalt pavements.

Regardless of the source, processing method, or mineralogy, aggregate must provide enough shear strength to resist repeated load applications. When a mass of aggregate is overloaded, a shear plane develops, and aggregate particles slide or *shear* with respect to each other (see Figure 1.1), resulting in permanent deformation. Along this plane, the *shear stress* exceeds the *shear strength* of the aggregate mass. Aggregate shear strength is critically important in HMA because it provides the mixture's primary rutting resistance.

Aggregate has relatively little cohesion. Thus, shear strength is

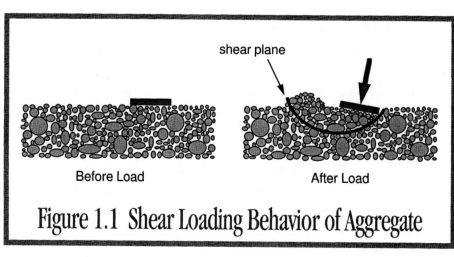

shear plane

Before Load

After Load

Figure 1.1 Shear Loading Behavior of Aggregate

primarily dependent on the resistance to movement, or internal friction, provided by the aggregates. Cubical, rough-textured aggregates provide more resistance than rounded, smooth-textured aggregates (see Figure 1.2). When a load is applied, aggregate also tends to be stronger because the load stress holds the aggregate tighter together and increases shear strength. Even though a cubical piece and rounded piece of aggregate may possess the same inherent strength, cubical aggregate particles tend to lock together resulting in a stronger mass of material. Rounded aggregate particles tend to slide by each other. The internal friction provides the ability of aggregate to interlock and create a mass that is almost as strong as the individual pieces.

To ensure a strong aggregate blend for HMA, aggregate properties

Cubical Aggregate   Rounded Aggregate

Figure 1.2 Aggregate Stone Skeleton

that enhance internal friction are typically specified. Normally, this is accomplished by specifying a certain percentage of crushed faces for the coarse portion of an aggregate blend. In addition, the amount of natural sand in a blend is often limited because natural sands tend to be rounded, with poor internal friction.

➤ ➤ Asphalt Mixture Behavior

When a wheel load is applied to a pavement, the primary stresses that are transmitted to the HMA are vertical compressive stress and shear stress within the asphalt layer, and horizontal tensile stress at the bottom of the asphalt layer. The HMA must be internally strong and resistant to compressive and shear stress to prevent permanent deformation within the mixture. In the same manner, the material must also have enough tensile strength to withstand tensile stress at the base of the asphalt layer to resist crack initiation, thus fatigue cracking after many load applications. The asphalt mixture must also resist stress imparted by rapidly decreasing temperatures and extremely cold temperatures.

While the individual properties of HMA components are important, asphalt mixture behavior is best explained by considering asphalt cement and mineral aggregate acting together. One way to understand asphalt mixture behavior is to consider the primary asphalt pavement distress types that engineers try to avoid: permanent deformation, fatigue cracking, and low temperature cracking. These are the distresses analyzed in Superpave.

Permanent deformation. *Permanent deformation* is characterized by a surface cross section that is no longer in its designed position. It is called permanent deformation because it represents an accumulation of small amounts of unrecoverable deformation that occur each time a load is applied. Wheel path rutting is the most common form of permanent deformation. While rutting can have many sources (e.g., underlying HMA weakened by moisture damage, abrasion, traffic densification), it has two principal causes.

In one case, rutting is caused by too much repeated stress being applied for the subgrade (or subbase or base) to withstand (see Figure 1.3). Although stiffer paving materials will partially reduce this type of rutting, it is normally considered a structural problem rather than a materials problem. Essentially, there is not enough pavement strength or thickness to reduce the applied stress to a tolerable level. It may also be caused by a pavement layer that has been unexpectedly weakened by the intrusion of moisture. The deformation occurs in the underlying layers rather than in the asphalt layers.

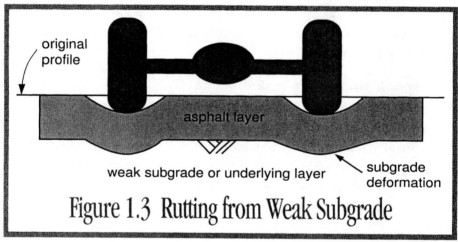

Figure 1.3 Rutting from Weak Subgrade

The type of rutting of most concern to asphalt mix designers is deformation in the asphalt layers. This rutting results from an asphalt mixture without enough shear strength to resist repeated heavy loads (see Figure 1.4). A weak mixture will accumulate small, but permanent, deformations with each truck pass, eventually forming a rut characterized by a downward and lateral movement of the mixture. The rutting may occur in the asphalt surface course; or, the rutting that shows on the surface may be caused by a weak underlying asphalt course.

Rutting of a unstable mixture typically occurs during the summer under higher pavement temperatures. While this might suggest that rutting is solely an asphalt cement problem, it is more correct to address rutting by considering the combined resistance of the mineral aggregate and asphalt cement.

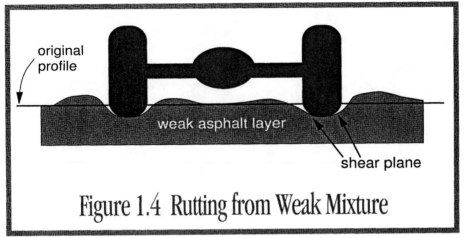

Figure 1.4 Rutting from Weak Mixture

Since rutting is an accumulation of very small permanent deformations, one way to increase mixture shear strength is to use a stiffer asphalt cement, that also behaves more like an elastic solid at high pavement temperatures. Then, when a load is applied, the asphalt cement will act like a rubber band and spring back to its original position rather than deforming.

Another way to increase the HMA shear strength is by selecting an aggregate that has a high degree of internal friction—one that is cubical, has a rough surface texture, and is graded to develop particle-to-particle contact. When a load is applied to the mixture, the aggregate particles lock tightly together and function more as a large, single, elastic stone. As with the asphalt cement, the aggregate will act like a rubber band and spring back to its original shape when unloaded. In that way, no permanent deformation accumulates.

Fatigue Cracking. Fatigue cracking occurs in asphalt pavements when the applied loads overstress the asphalt materials, causing cracks to form. An early sign of fatigue cracking is intermittent longitudinal cracks in the traffic wheel path. Fatigue cracking is progressive because at some point the initial cracks will join, causing even more cracks to form. An advanced stage of fatigue cracking is called alligator cracking, characterized by transverse cracks connecting the longitudinal cracks (see Figure 1.5). In extreme cases, a pothole forms when pavement pieces become dislodged by traffic.

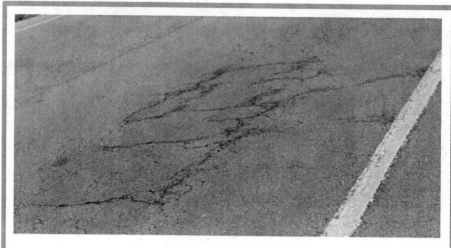

Figure 1.5 Fatigue (Alligator) Cracking

Fatigue cracking is usually caused by a number of factors occurring simultaneously. Obviously, repeated heavy loads must be present. Thin pavements or those with weak underlying layers are prone to high deflections under heavy wheel loads. High deflections (repeated bending and flexing) cause horizontal tensile stress at the bottom of the asphalt layer, leading to fatigue cracking. Poor drainage, poor construction, and/or an underdesigned pavement can contribute to this problem.

Often, fatigue cracking is merely a sign that a pavement has sustained the design number of load applications, in which case the pavement is simply in need of planned rehabilitation. If fatigue cracking occurs at the end of the design period, it would be considered a natural progression of the pavement design strategy. If the observed cracking occurs much sooner than the design period, it may be a sign that traffic loads were underestimated. Consequently, the best ways to overcome fatigue cracking are:

▲ Adequately account for the number of heavy loads during design.
▲ Use thicker pavements.
▲ Keep the subgrade dry.
▲ Use pavement materials not excessively weakened by moisture.
▲ Use HMA that is resilient enough to withstand normal deflections.

Only the last item, selection of resilient materials, can be strictly addressed using materials selection and mix design. The HMA must have enough tensile strength to withstand the applied tensile stress at the base of the asphalt layer, and be resilient enough to withstand repeated load applications without cracking. Thus, HMA must be designed to behave like a soft elastic material when loaded in tension to overcome fatigue cracking. This is accomplished by placing an upper limit on the asphalt cement's stiffness properties, since the tensile behavior of HMA is strongly influenced by the asphalt cement. In effect, soft asphalts have better fatigue properties than hard asphalts.

Low Temperature Cracking. Low temperature cracking is caused by adverse environmental conditions rather than by applied traffic loads. It is characterized by intermittent transverse cracks that occur at a surprisingly consistent spacing (Figure 1.6).

Low temperature cracks form when an asphalt pavement layer shrinks in cold weather. As the pavement shrinks, tensile stresses build within the layer. At some point along the pavement, the tensile stress exceeds the tensile strength and the asphalt layer cracks. Low temperature cracks occur primarily from a single cycle of low temperature, but can develop from repeated low temperature cycles.

The asphalt binder plays a key role in low temperature cracking. In general, hard asphalt binders are more prone to low temperature cracking than soft asphalt binders are. Asphalt binders that are excessively aged, because they are unduly prone to oxidation and/or contained in a mixture constructed with too many air voids, are more prone to low temperature cracking. Thus, to overcome low temperature cracking, engineers must use a soft binder that is not overly prone to aging, and control the in-place air void content of the pavement so that the binder does not become excessively oxidized.

Figure 1.6 Low Temperature Cracking

➤➤ Penetration and Viscosity Graded Specifications

Because asphalt is chemically complex, specifications for its use have been developed around physical property tests. Traditional specifications use such tests as penetration, viscosity, and ductility. These physical property tests are performed at standard test temperatures, and the test results are used to determine if the material meets the specification criteria.

However, there are limitations to what traditional test procedures provide in terms of results. Many of these tests are empirical, meaning that field experience is required before the test results yield meaningful information. The penetration test represents the stiffness of the asphalt, but any relationship between asphalt penetration and performance has to be gained by experience.

Another limitation to empirical tests and traditional specifications is that the tests do not give information for the entire range of typical pavement temperatures. Although viscosity is a fundamental measure of flow, it only provides information about higher temperature viscous behavior. The standard test temperatures are 60°C and 135°C. Lower temperature elastic behavior cannot be realistically determined from this data to completely predict performance. As well, penetration describes only the consistency at a medium temperature (25°C). No low temperature properties are directly measured in the current grading systems.

When using viscosity and penetration grading in asphalt specifications, asphalts with very different temperature and performance characteristics can be classified the same. As an example, Figure 2.1 shows three asphalts that have the same viscosity grade. They are within the specified viscosity limits at 60°C, have the minimum penetration at 25°C, and reach the minimum viscosity at 135°C. While Asphalt A and B display the same temperature dependency, they have much different consistency at all temperatures. Asphalt A and C have the same consistency at low temperatures, but remarkably different high temperature consistency. Asphalt B has the same consistency at 60°C, but shares no other similarities with Asphalt C. Because these asphalts meet the same grade specifications, one might erroneously expect the same characteristics during construction and the same performance during hot and cold weather conditions.

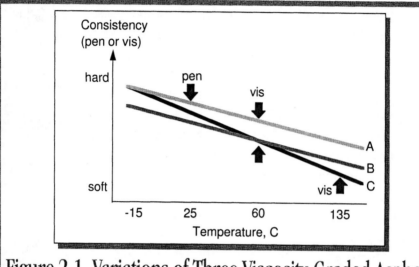

Figure 2.1 Variations of Three Viscosity-Graded Asphalts

Because of these deficiencies, many state highway agencies have amended standard test procedures and specifications to better suit local conditions. In some parts of the U.S., this proliferation of tests and specifications has caused problems for asphalt suppliers wanting to sell the same asphalt grades in several states. Often, states with very similar performance conditions and materials will specify significantly different asphalt cement requirements. Also, tests in traditional asphalt specifications are performed on unaged or *tank* asphalt and on asphalt artificially aged to simulate construction aging. However, no tests are performed on asphalt cements that have been treated to simulate in-service aging.

➤ ➤ Superpave Performance Graded Asphalt Binder Specification

The Superpave asphalt binder specification differs from other asphalt specifications in that the tests used measure physical properties that can be directly related to field performance by engineering principles. The

Superpave binder specification requires a new set of test equipment and procedures. It is called a "binder" specification because it is intended for both modified and unmodified asphalt.

A unique feature of the Superpave binder specification is that instead of performing a test at a constant temperature and varying the specified value, the specified value is constant and the test temperature at which this value must be achieved is varied. As an example, consider two construction projects—one at the Equator and one at the Arctic Circle. Good asphalt performance is expected in both locations, but the temperature conditions under which good binder performance is achieved are vastly different.

Performance graded (PG) binders are defined by a term such as PG 64-22. The first number, 64, is the *high temperature grade*. This means that the binder possesses adequate physical properties up to at least 64°C. This would correspond with the high pavement temperature in the climate in which the binder is expected to serve. Likewise, the second number (-22) is the *low temperature grade* and means that the binder possesses adequate physical properties in pavements down to at least -22°C.

The central theme of the Superpave binder specification is its reliance on testing asphalt binders in conditions that simulate the three critical stages during the binder's life. Tests performed on the original asphalt represent the first stage of transport, storage, and handling. The second stage represents the asphalt during mix production and construction, and is simulated in the specification by aging the binder in a rolling thin film oven. This procedure exposes thin binder films to heat and air and approximates the exposure of asphalt during hot mixing, hauling and laydown conditions. The third stage occurs as the binder ages over a long period as part of the hot mix asphalt pavement layer. This stage is simulated in the specification by the pressure-aging vessel. This procedure exposes binder samples to heat and pressure conditions that simulate years of in-service aging in a pavement.

The Superpave binder specification is continually being evaluated by both AASHTO and ASTM. Table 2.1 lists the Superpave binder test equipment and a brief description of its use in the specification. Details on how to conduct tests for the Superpave binder specification, and the test results' relationship to field performance, are presented in *Performance Graded Asphalt Binder Specification and Testing,* Superpave Series No. 1 (SP-1).

Table 2.1 Superpave Binder Test Equipment

Equipment	Purpose
Rolling Thin Film Oven (RTFO) Pressure Aging Vessel (PAV)	Simulate binder aging (hardening) characteristics
Dynamic Shear Rheometer (DSR)	Measure binder properties at high and intermediate temperatures
Rotational Viscometer (RV)	Measure binder properties at high temperatures
Bending Beam Rheometer (BBR) Direct Tension Tester (DTT)	Measure binder properties at low temperatures

Asphalt Mixture Design Procedures

➤➤ Marshall and Hveem Methods

Most agencies traditionally use the Marshall mix design method. Developed by Bruce Marshall of the Mississippi State Highway Department, the U.S. Army Corps of Engineers refined and added certain features to Marshall's approach and it was formalized as ASTM D1559 and AASHTO T245. The Marshall method entails a laboratory experiment aimed at developing a suitable asphalt mixture using stability/flow and density/voids analyses.

One advantage of the Marshall method is its attention to density and voids properties of asphalt mixtures. This analysis ensures the proper volumetric proportions of mixture materials for achieving a durable HMA. Another advantage is that the required equipment is relatively inexpensive and portable, and thus, lends itself to remote quality control operations. However, many engineers believe that the impact compaction used with the Marshall method does not simulate mixture densification as it occurs in a real pavement. Furthermore, Marshall stability does not adequately estimate the shear strength of HMA. These two situations make it difficult to assure the rutting resistance of the designed mixture. Consequently, there has been a growing feeling among asphalt technologists that the Marshall method has outlived its usefulness for modern asphalt mixture design.

The Hveem mix design procedure was developed by Francis Hveem of the California Department of Transportation. Hveem and others refined the procedure, which is detailed in ASTM D1560 and ASTM D1561. The Hveem method is not commonly used for HMA outside the western United States.

The Hveem method also entails a density/voids and stability analysis. The mixture's resistance to swell in the presence of water is also determined. The Hveem method has two primary advantages. First, the kneading method of laboratory compaction is thought to better simulate the densification characteristics of HMA in a real pavement. Second, Hveem stability is a direct measurement of the internal friction component of shear strength. It measures the ability of a test specimen to resist lateral displacement from application of a vertical load.

A disadvantage of the Hveem procedure is that the testing equipment is somewhat expensive and not very portable. Furthermore, some important mixture volumetric properties that are related to mix durability are not routinely determined as part of the Hveem procedure. Some engineers believe that the method of selecting asphalt content in the Hveem method is too subjective and may result in non-durable HMA with too little asphalt.

To augment the standard Marshall and Hveem methods, agencies have increasingly adopted laboratory design procedures, methods and/or systems that they have found suitable for their conditions. The Georgia loaded wheel tester is an example of equipment used to replace or supplement procedures in other design systems. The advantage of these systems is that agencies can develop very clear criteria, backed up by performance data from real pavements. However, agencies also have to conduct many experiments to achieve this experience. Even then the experience is only applicable to the materials and environmental conditions tested. New products and materials require additional experimentation.

➤ ➤ Superpave Asphalt Mixture Design Procedures

Individual steps are typically used to select asphalt and aggregate materials and conduct mix design procedures to combine the materials. The Superpave mix design system integrates material selection and mix design into procedures based on the project's climate and design traffic.

The Superpave asphalt binder specification was discussed earlier in this chapter and the aggregate and mix design specifications are discussed here.

Mineral Aggregates. Mineral aggregate properties are obviously important to asphalt mixture performance. However, the Marshall and Hveem mix design methods do not incorporate aggregate criteria into their procedures. Conversely, aggregate criteria are directly incorporated into Superpave mix design procedures. While no new aggregate test procedures were developed, existing procedures were refined to fit within the Superpave system.

Two types of aggregate properties are specified in the Superpave system: *consensus* properties and *source* properties. Consensus properties are those that SHRP researchers believed were critical in achieving high performance HMA. These properties must be met at various levels depending on traffic volume and position within the pavement. High traffic levels and surface mixtures (i.e., shallow pavement position) require more strict values for consensus properties.

Many agencies already use these properties as quality requirements for aggregates used in HMA. The consensus properties identified in Superpave are:

▲ Coarse Aggregate Angularity
▲ Fine Aggregate Angularity
▲ Flat and Elongated Particles
▲ Clay Content

By specifying coarse and fine aggregate angularity, Superpave seeks to achieve HMA with a high degree of internal friction and thus, high shear strength for rutting resistance. Limiting elongated pieces ensures that the HMA will not be as susceptible to aggregate breakage during handling and construction and under traffic. Limiting the amount of clay enhances the adhesive bond between asphalt binder and the aggregate.

Source properties are those that agencies often use to qualify local sources of aggregate. While these properties are important, critical values are not specified in the Superpave specification since they are source specific. The source properties identified in Superpave are:

▲ Toughness
▲ Soundness
▲ Deleterious Materials

Toughness is measured by the LA abrasion test. Soundness is measured by the sodium or magnesium sulfate soundness test. Deleterious materials are measured by the clay lumps and friable particles test.

To specify aggregate gradation, Superpave uses the 0.45-power gradation chart with gradation control limits and a restricted zone to develop a design aggregate structure. Superpave requires that the *design aggregate structure* pass between gradation control points while recommending that the gradation restricted zone be avoided. The restricted zone is used by Superpave to avoid mixtures that have a high proportion of fine sand relative to total sand and, to avoid gradations that follow the maximum density line, which do not normally have adequate voids in the mineral aggregate (VMA). In many instances, the restricted zone will discourage the use of fine natural sand in an aggregate blend and encourage the use of clean manufactured sand.

The design aggregate structure approach ensures that the aggregate will develop a strong stone skeleton to enhance resistance to permanent deformation while allowing for sufficient void space to enhance mixture durability.

Superpave Mixture Design. One of the key features in Superpave mix design is the change in laboratory compaction methods. Laboratory compaction is accomplished using a Superpave gyratory compactor (SGC). The SGC shares some traits with existing gyratory compactors, but it has completely new operational characteristics. While its main purpose is to compact test specimens, the SGC can provide information about the compactability of the particular mixture by capturing data during compaction. The SGC can be used to design mixtures that do not exhibit classic tender mix behavior and do not densify to dangerously low air void contents under traffic action.

Superpave mix design procedures depend on the traffic level of the pavement for which the HMA is being designed. The procedure, called volumetric mix design, can be used for all pavement projects, and it entails compacting test specimens using the SGC and selecting asphalt content on the basis of volumetric design requirements.

The performance of HMA immediately after construction is influenced by mixture properties resulting from hot mixing and compaction. Consequently, a short term aging protocol was incorporated into the Superpave system. This was accomplished by requiring that samples, prior to compaction by the SGC, be oven aged to simulate the delays that occur in actual construction activities. Originally, the aging period and temperature were four hours at 135°C, but has been revised to a two-hour period at the compaction temperature.

The performance based tests and performance prediction models for HMA are important developments from the SHRP asphalt research. At the time of this printing however, they are still being developed. Output from mixture performance tests will be designed to make detailed predictions of actual pavement performance. These tools will allow an engineer to estimate the performance life of a prospective HMA in terms of equivalent single axle loads (ESAL).

CHAPTER 3

Asphalt Binder Superpave uses an improved system for testing, specifying, and selecting asphalt binders. The Superpave binder specification, AASHTO MP1, is unique in that it is performance based and that binders are selected on the basis of the climate and traffic in which they are intended to serve. Discussion in this manual is centered on selecting the correct Performance Grade (PG) for a particular application. For more details on Superpave binder test equipment, procedures, and criteria, see the Asphalt Institutes' *Performance Graded Asphalt Binder Specification and Testing* (SP-1).

The physical property requirements within the specification are constant among all PG grades. What differentiates the various binder grades is temperature at which the requirements must be met. For example, a binder classified as a PG 64-22 means that the binder must meet high-temperature physical property requirements, at least up to a temperature of 64°C, and low temperature physical properties must be met at least down to -22°C. These physical properties are directly related to field performance, so the greater the first (high) temperature value is, the more resistant the binder should be to high temperature distress such as rutting or shoving. Likewise, the lower the second (low) PG temperature value is, the more resistant the binder should be to low temperature cracking. The high and low temperature designations extend in both directions as far as necessary in six-degree increments, making the number of possible grades almost unlimited. The more common paving grades used in the US are PG 64-22, PG 70-22, PG 76-22, PG 58-22, PG 64-28, PG 58-28, and PG 52-34.

Specifiers select a binder grade based on environment (historical temperature data), traffic conditions, and the desired reliability factor. Most agencies specify, in the project contract, the binder grade to be used. If the designer is responsible for selecting the PG grade, two software programs are available to aid in choosing the appropriate grade. First, there is the *LTPP Bind* software developed by, and available through, the Long-term Pavement Performance Program (LTPP) of the FHWA Turner Fairbanks Highway Research Center. Second, there is the AASHTO Superpave program, a registered product of the *AASHTO ROADWare* Family. While *LTPP Bind* deals strictly with PG binder selection, *AASHTO Superpave* assists in the entire Superpave mix design process – along with providing quality control and quality assurance guidance during production. One module of *AASHTO Superpave* deals with PG binder selection.

MATERIALS SELECTION

These programs provide a database of temperature information from over 7000 weather stations in the US and Canada to allow users to select binder grades for the climate at a particular project location. For each year that the weather stations have been in operation, the hottest seven-day period was identified, and the average maximum air temperature for this seven-day period was calculated. For all the years of operation, the mean and standard deviation of the seven-day average maximum air temperature was calculated. Similarly, the one-day minimum air temperature of each year was identified, and the mean and standard deviation were calculated. Weather stations with less than 20 years of data were not used.

However, the design temperatures to be used for selecting asphalt binder grades are the pavement temperatures, not the air temperatures. Therefore, air temperatures from the weather station database must be converted into pavement temperatures. For surface layers, Superpave defines the locations for the high pavement design temperature at a depth 20 mm below the pavement surface, and the low pavement design temperature at the pavement surface.

Using theoretical analyses of actual conditions performed with models for net heat flow and energy balance, and assuming typical values for solar absorption (0.90), radiation transmission through air (0.81), atmospheric radiation (0.70), and wind speed (4.5 m/sec), this equation was developed to convert the seven-day high air temperature to the high pavement design temperature:

$$T_{20mm} = (T_{air} - 0.00618 \, Lat^2 + 0.2289 \, Lat + 42.2) (0.9545) - 17.78$$

where T_{20mm} = high pavement design temperature at a depth of 20 mm

T_{air} = seven-day average high air temperature, °C

Lat = the geographical latitude of the project in degrees.

The preferred method of determining the low pavement design temperature in Superpave is to utilize the *LTPP Bind* software. Prior to the availability of this software, there were two possible ways to establish this temperature. The first way was to simply assume that the low pavement design temperature was the same as the low air temperature. This method was originally recommended by SHRP researchers. This is a very conservative assumption because pavement temperature is almost always warmer than air temperature in cold weather. The second method uses an equation developed by Canadian SHRP researchers:

$$T_{min} = 0.859 T_{air} + 1.7°C$$

where T_{min} = minimum pavement design temperature in °C,

T_{air} = minimum air temperature in average year in °C.

The Superpave system allows the designers to use reliability measurements to assign a degree of design risk to the high and low pavement temperatures used in selecting the binder grade. Reliability is the percent probability in a single year that the actual temperature (one-day low or seven-day high) will not exceed the design temperature. A higher reliability means lower risk. For example, consider summer air temperatures in Cleveland, Ohio, which has a mean seven-day maximum of 32°C and a standard deviation of 2°C. In an average year, there is a 50 percent chance that the seven-day maximum air temperature will exceed 32°C. However, assuming a normal statistical frequency distribution, there is only a two percent chance that the seven-day maximum will exceed 36°C (mean plus two standard deviations); therefore, as shown in Figure 3.1, a design air temperature of 36°C will provide 98 percent reliability.

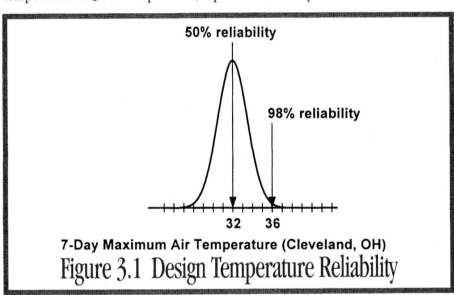

50% reliability

98% reliability

32   36

7-Day Maximum Air Temperature (Cleveland, OH)

Figure 3.1 Design Temperature Reliability

➤➤ Air Temperature Selection

Continuing the example, assume that an asphalt mixture is to be designed for Cleveland. Figure 3.2 shows the statistical variation of the two design air temperatures. In a normal summer, the average seven-day maximum air temperature is 32°C and in a "very hot" summer, this average may reach 36°C. Using the same approach for winter conditions, Cleveland has a one-day minimum air temperature of -21°C with a standard deviation of 4°C. Consequently, in an average winter, the coldest temperature is -21°C.

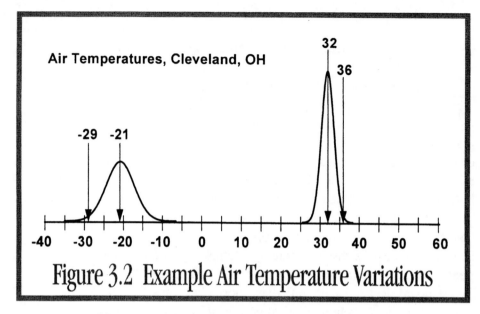

Figure 3.2 Example Air Temperature Variations

For a "very cold" winter, the temperature may reach -29°C. The standard deviations show there is more variation in the one-day low temperatures than the seven-day average high temperatures.

➤➤ Pavement Temperature Selection

Continuing the example, for a surface course in Cleveland the design pavement temperatures are about 52°C and -16°C for 50 percent reliability and about 56°C and -23°C for 98 percent reliability (mean plus two standard deviations). Figure 3.3 graphically represents the statistical variation of the two pavement temperatures.

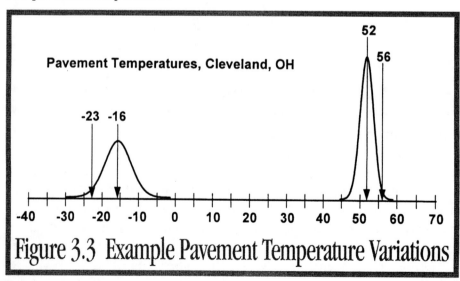

Figure 3.3 Example Pavement Temperature Variations

➤➤ Binder Grade Selection Based on Pavement Temperatures

To achieve a reliability of at least 50 percent and provide for an average maximum pavement temperature of at least 52°C, the standard high temperature grade, PG 52, happens to match the design temperature, 52°C. Using the same reasoning, the standard low temperature grade to attain 50 percent reliability is a PG -16, which again happens to match the design reliability. As shown in Figure 3.4, to obtain at least 98% reliability, it is then necessary to select a standard high temperature grade of PG 58 to protect above 56°C and a standard low temperature grade of PG -28 to protect below -23°C. In both the high and low temperature cases of the PG 58-28 binder grade, the actual reliability exceeds 99 percent because of the "rounding up" caused by the six degree difference between standard grades. This "rounding up" introduces conservatism into the binder selection process.

Another source of conservatism occurs when considering the actual asphalt binder physical properties. The specific binder may possess the actual properties of a PG 60-24, but it will nevertheless be classified to a standard grade of PG 58-22. The net result is that a significant factor of safety is included in the binder selection scheme. For example, it is possible that the PG 52-16 binder, selected previously for a minimum of 50 percent reliability for Cleveland, may actually have been graded as a PG 56-20, had such a grade existed. Users of the PG grading system for binder selection should recognize that considerable safeguards are already included in the process. Because of these factors, it may not be necessary or cost-effective to require indiscriminately high values of reliability or abnormally conservative high or low temperature grades.

The Superpave computer programs perform all of these calculations based on minimal user input. For any location, the user can enter a minimum reliability and the software will calculate the required asphalt binder grade. Alternatively, the user can specify a desired asphalt binder grade and Superpave will calculate the reliability. When deciding the asphalt binder PG grades to specify for their climatic and loading conditions, agencies are faced with engineering management decisions. They will have to decide the level of reliability to be used. Depending on the policy established by each individual agency, the selected reliability may be a function of road classification, traffic level, binder cost and availiability, and other factors.

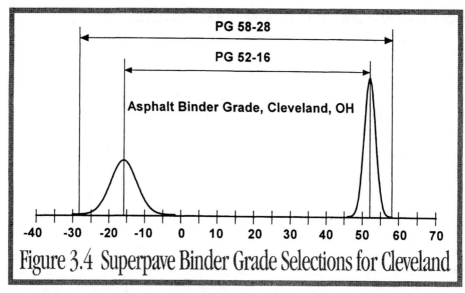

Figure 3.4 Superpave Binder Grade Selections for Cleveland

➤➤ Adjusting Binder Grade Selection For Traffic Speed and Loading (Grade Bumping)

The asphalt binder selection procedure described is the basic procedure for typical highway loading conditions. Under these conditions, it is assumed that the pavement is subjected to a design number of fast, transient loads. For the high temperature design situation, controlled by specified properties relating to permanent deformation, the speed of loading has an additional effect on performance.

Superpave requires an additional shift in the selected high temperature binder grade for slow and standing load applications. For slow moving design loads, the binder would be selected one high temperature grade higher, such as a PG 64 instead of a PG 58. For standing design loads, the binder would be selected two high temperature grades higher, such as a PG 70 instead of a PG 58.

Also, an additional shift is needed for extraordinarily high numbers of heavy traffic loads. If the design traffic is expected to be between 10,000,000 and 30,000,000 equivalent single axle loads (ESAL), then the engineer is encouraged to consider selecting one high temperature binder grade higher than the selection based on climate. If the design traffic is expected to exceed 30,000,000 ESAL, then the binder is required to be selected one high temperature grade higher than the selection based on climate. This practice of adjusting high temperature grades for traffic loading and speed is sometimes called "grade-bumping". Table 3.1 summarizes AASHTO's grade-bumping policy as it is presented in MP2 of the AASHTO Provisional Standards.

It should be emphasized that proper or conservative binder selection does not guarantee total pavement performance. Fatigue cracking performance is greatly affected by the pavement structure and traffic. Permanent deformation or rutting is directly a function of the shear strength of the mix, which is greatly influenced by aggregate properties. Pavement low temperature cracking correlates most significantly to the binder properties. Engineers should try to achieve a balance among the many factors when selecting binders.

Table 3.1 Binder Selection on the Basis of Traffic Speed and Traffic Level

Design ESALs[1] (million)	Adjustment to Binder PG Grade[5] Traffic Load Rate		
	Standing[2]	Slow[3]	Standard[4]
< 0.3	_[6]		
0.3 to < 3	2	1	
3 to < 10	2	1	
10 to < 30	2	1	_[6]
≥ 30	2	1	1

1. Design ESALs are the anticipated project traffic level expected on the design lane over a 20 year period. Regardless of the actual design life of the roadway determine the design ESALs for 20 years and choose the appropriate N_{design} level.

2. Standing Traffic - where the average traffic speed is less than 20 km/h.

3. Slow Traffic - where the average traffic speed ranges from 20 to 70 km/h.

4. Standard Traffic - where the average traffic speed is greater than 70 km/h.

5. Increase the high temperature grade by the number of grade equivalents indicated (one grade equivalent to 6°C). Do not adjust the low temperature grade.

6. Consideration should be given to increasing the high temperature grade by one grade equivalent.

Practically, performance graded binders stiffer than PG 82-XX should be avoided. In cases where the required adjustment to the high temperature binder grade would result in a grade higher than a PG 82, consideration shold be given to specifying a PG 82-XX and increasing the design ESALs by one level (e.g., 10 to < 30 million increased to ≥ 30 million).

Mineral Aggregate

SHRP researchers surveyed pavement experts to determine which aggregate properties were most important. There was general agreement that aggregate properties played the integral role in overcoming permanent deformation. Fatigue cracking and low-temperature cracking were less affected by aggregate characteristics. SHRP researchers used these survey results to identify two categories of aggregate properties that needed to be used in the Superpave system: *consensus* properties and *source* properties. Additionally, a new way of specifying the design aggregate gradation was developed.

➤➤ Consensus Aggregate Properties

The pavement experts agreed that certain aggregate characteristics were critical to well-performing HMA. These characteristics were called "consensus properties" because there was wide aggreement in their use and specified values. As discussed in chapter 2, the consensus properties are *coarse aggregate angularity, fine aggregate angularity, flat and elongated particles and clay content.*

The criteria for these consensus aggregate properties are based on traffic level and position within the pavement structure. Materials near the pavement surface subjected to high traffic levels require more stringent consensus properties. The criteria are intended to be applied to a proposed aggregate blend rather than individual components. However, many agencies currently apply such requirements to individual aggregates so undesirable components can be identified. The consensus properties are detailed below (also see Table 3.2).

Coarse Aggregate Angularity. This property ensures a high degree of aggregate internal friction and rutting resistance. It is defined as the percentage (by mass) of aggregates larger than 4.75 mm with one or more fractured faces. The test method specified by Superpave is ASTM D5821, *Test Method for Determining the Percentage of Fractured Faces in Coarse Aggregate.* Table 3.2 gives the required minimum values for coarse aggregate angularity as a function of traffic level and position within the pavement.

Table 3.2 Superpave Aggregate Consensus Property Requirements

Design ESALs[1] (million)	Coarse Aggregate Angularity (Percent), minimum		Uncompacted Void Content of Fine Aggregate (Percent), minimum		Sand Equivalent (Percent), minimum	Flat and Elongated[3] (Percent), maximum
	≤ 100 mm	> 100 mm	≤ 100 mm	> 100 mm		
< 0.3	55/-	-/-	-	-	40	-
0.3 to< 3	75/-	50/-	40	40	40	10
3 to < 10	85/80[2]	60/-	45	40	45	10
10 to < 30	95/90	80/75	45	40	45	10
≥ 30	100/100	100/100	45	45	50	10

1. Design ESALs are the anticipated project traffic level expected on the design lane over a 20 year period. Regardless of the actual design life of the roadway determine the design ESALs for 20 years and choose the appropriate N_{design} level.

2. 85/80 denotes that 85% of the coarse aggregate has one fractued face and 80% has two or more fractured faces.

3. Criterion based upon a 5:1 maximum-to-minimum ratio.

(If less than 25% of a layer is within 100 mm of the surface, the layer may be considered to be below 100 mm for mixture design purposes.)

Fine Aggregate Angularity. This property ensures a high degree of fine aggregate internal friction and rutting resistance. It is defined as the percent air voids present in loosely compacted aggregates smaller than 2.36 mm. The test method specified by Superpave is AASHTO T304, *Uncompacted Void Content of Fine Aggregate.* This property is influenced by particle shape, surface texture, and grading. Higher void contents mean more fractured faces.

In the test procedure, a sample of fine, washed and dried aggregate is poured into a small calibrated cylinder through a standard funnel (Figure 3.5). By measuring the mass of fine aggregate (W) in the filled cylinder of known volume (V), the void content can be calculated as the difference between the cylinder volume and fine aggregate volume collected in the cylinder. The fine aggregate bulk specific gravity (G_{sb}) is used to compute the fine aggregate volume.

Table 3.2 gives the required minimum values for fine aggregate angularity (Uncompacted Void Content of Fine Aggregate) as a function of traffic level and position within the pavement.

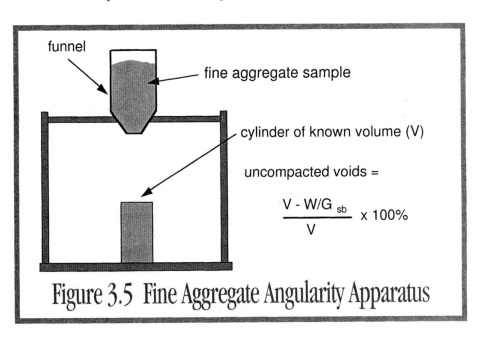

funnel

fine aggregate sample

cylinder of known volume (V)

uncompacted voids =

$$\frac{V - W/G_{sb}}{V} \times 100\%$$

Figure 3.5 Fine Aggregate Angularity Apparatus

Flat and Elongated Particles. This characteristic is the percentage by mass of coarse aggregates that have a maximum to minimum dimension ratio greater than five. Flat and elongated particles are undesirable because they have a tendency to break during construction and under traffic. The test procedure used is ASTM D4791, *Flat or Elongated Particles in Coarse Aggregate* and it is performed on coarse aggregate larger than 4.75 mm.

The procedure uses a proportional caliper device (Figure 3.6) to measure the dimensional ratio of a representative sample of aggregate particles. In Figure 3.6, the aggregate particle is first placed with its largest dimension between the swinging arm and fixed post at position A. The swinging arm then remains stationary while the aggregate is placed between the swinging arm and the fixed post at position B. If the aggregate does not fill this gap, then it is counted as a flat or elongated particle.

The required maximum values for flat and elongated particles in coarse aggregate are given in Table 3.2.

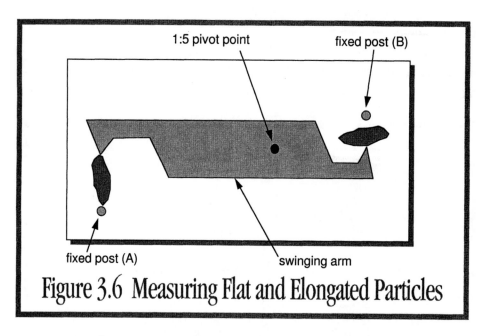

Figure 3.6 Measuring Flat and Elongated Particles

Clay Content (Sand Equivalent). Clay content is the percentage of clay material contained in the aggregate fraction that is finer than a 4.75-mm sieve. It is measured by AASHTO T176, *Plastic Fines in Graded Aggregates and Soils by Use of the Sand Equivalent Test* (ASTM D2419).

A sample of fine aggregate is mixed with a flocculating solution in a graduated cylinder and agitated to loosen clayey fines present in and coating the aggregate. The flocculating solution forces the clayey material into suspension above the granular aggregate. After a settling period, the cylinder height of suspended clay and settled sand is measured (Figure 3.7). The *sand equivalent* value is computed as the ratio of the sand to clay height readings, expressed as a percentage.

The allowable clay content values for fine aggregate, expressed as a minimum percentage of sand equivalent, are given in Table 3.2.

➤ ➤ Source Aggregate Properties

In addition to the consensus aggregate properties, SHRP researchers believed that certain other aggregate characteristics were critical. However, critical values of these properties could not be reached by consensus because needed values were source specific. Consequently, a set of *source properties* was recommended.

Specified values are established by local agencies. While these properties are relevant during the mix design process, they may also be used as source acceptance control. Those properties are *toughness, soundness,* and *deleterious materials.*

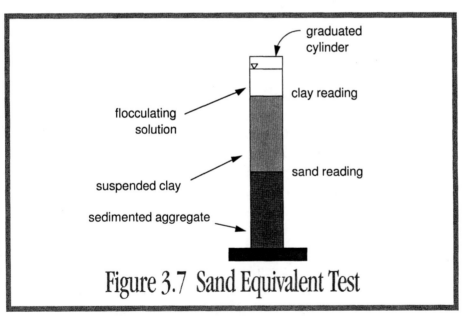

Figure 3.7 Sand Equivalent Test

Toughness. Toughness is the percent loss of material from an aggregate blend during the Los Angeles Abrasion test (AASHTO T96 or ASTM C131 or C535). This test estimates the resistance of coarse aggregate to abrasion and mechanical degradation during handling, construction and in-service. It is performed by subjecting the coarse aggregate, usually larger than 2.36 mm, to impact and grinding by steel spheres. The test result is the mass percentage of coarse material lost during the test due to the mechanical degradation. Maximum loss values typically range from 35 to 45 percent.

Soundness. Soundness is the percent loss of material from an aggregate blend during the sodium or magnesium sulfate soundness test (AASHTO T104 or ASTM C88). This test estimates the resistance of aggregate to in-service weathering. It can be performed on both coarse and fine aggregate. The test is performed by exposing an aggregate sample to repeated immersions in saturated solutions of sodium or magnesium sulfate followed by oven drying. One immersion and drying is considered one soundness cycle. During the drying phase, salts precipitate in the permeable void space of the aggregate. Upon re-immersion the salt rehydrates and exerts internal expansive forces that simulate the expansive forces of freezing water.

The test result is total percent loss over various sieve intervals for a required number of cycles. Maximum loss values typically range from 10 to 20 percent for five cycles.

Deleterious Materials. Deleterious materials are defined as the mass percentage of contaminants such as clay lumps, shale, wood, mica, and coal in the blended aggregate (AASHTO T112 or ASTM C142). The analysis can be performed on both coarse and fine aggregate. The test is performed by wet sieving aggregate size fractions over specified sieves. The mass percentage of material lost as a result of wet sieving is reported as the percent of clay lumps and friable particles.

A wide range of criteria for maximum allowable percentage of deleterious particles exists. Values range from as little as 0.2 percent to as high as 10 percent, depending on the exact composition of the contaminant.

➤ ➤ Gradation

To specify gradation, Superpave modifies an approach already used by some agencies. The 0.45-power gradation chart is used to define a permissible gradation. The "point 45 power" chart uses a unique graphing technique to show the cumulative particle size distribution of an aggregate blend. The ordinate (vertical axis) of the chart is percent passing. The abscissa (horizontal axis) is an arithmetic scale of sieve size in millimeters, raised to the 0.45 power. Figure 3.8 illustrates how the abscissa is scaled.

In this example, the 4.75-mm sieve is plotted at 2.02, which is the sieve size of 4.75 mm raised to the 0.45 power. Traditionally, 0.45-power charts do not show arithmetic abscissa labels such as those in Figure 3.8. Instead, the scale is marked with the actual sieve size, as in Figure 3.9.

An important feature of the 0.45 power chart is the maximum density gradation. The maximum density line plots as a straight line from the

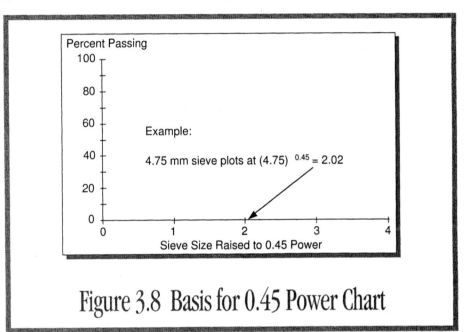

Figure 3.8 Basis for 0.45 Power Chart

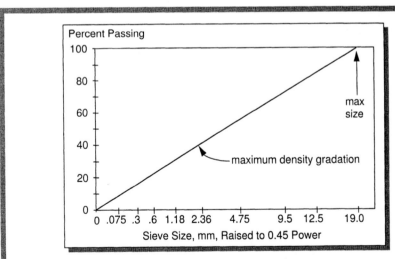

Figure 3.9 Maximum Density Gradation for 19mm Maximum Size

maximum aggregate size to the origin. Below are the Superpave definitions for maximum and nominal maximum size:

Maximum Size. One sieve size larger than the nominal maximum size.
Nominal Maximum Size. One sieve size larger than the first sieve to retain more than 10 percent.

The maximum density line (Figure 3.9) represents a gradation where the aggregate particles fit together in their densest possible arrangement. Figure 3.9 shows a 0.45-power gradation chart with a maximum density line for a 19.0-mm maximum size aggregate (12.5-mm nominal maximum aggregate size).

To specify aggregate gradation, two additional features have been added to the 0.45-power chart—control points and a restricted zone.

Control Points. Control points function as master ranges through which gradations must pass. Control points are placed at the nominal maximum size, an intermediate size (2.36 mm), and the smallest size (0.075 mm). The control point limits vary depending on the nominal maximum aggregate size of the design mixture.

Restricted Zone. The restricted zone resides along the maximum density gradation between the intermediate size (either 4.75 mm or 2.36 mm) and the 0.3-mm size. Figure 3.10 shows the control points and restricted zone for a 12.5-mm Superpave mixture (12.5-mm nominal maximum and 19.0-mm maximum size).

The restricted zone forms a band through which it is generally recommended that the gradation not pass. Gradations that pass through the restricted zone from below the zone have often been called "humped gradations" because of the characteristic hump in the grading curve that passes through the restricted zone.

In most cases, a humped gradation indicates an over-sanded mixture and/or a mixture that possesses too much fine sand in relation to total sand. This gradation often results in tender mix behavior, which is manifested by compaction problems during construction. These mixtures may also offer reduced resistance to permanent deformation (rutting) during their performance life.

The restricted zone prevents a gradation from following the maximum density line in the fine aggregate sieves. Gradations that follow the maximum density line often have inadequate VMA to allow room for sufficient asphalt for durability. These gradations are typically very sensitive to asphalt content and can easily become plastic with even minor variations in asphalt content.

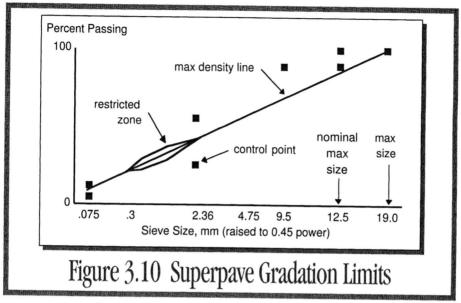

Figure 3.10 Superpave Gradation Limits

While Superpave originally recommended that gradations pass below the restricted zone, it is not a requirement. Several highway agencies successfully use gradings passing above the restricted zone. Experience has shown that some gradations passing through the restricted zone perform satisfactorily. Before using such gradations, it is recommended that experience or testing be evaluated to determine if the particular aggregate structure performs satisfactorily (adequate VMA, non-tender mix behavior, etc.).

The term used to describe the distribution of aggregate particle sizes is the *design aggregate structure*. A design aggregate structure that lies between the control points meets the Superpave gradation requirements.

Superpave defines five mixture gradations by their nominal maximum aggregate size (Table 3.3). Appendix A shows numerical gradation limits for the five Superpave mixtures.

Table 3.3 Superpave Mixture Gradations		
Superpave Designation	Nominal Maximum Size, mm	Maximum Size, mm
37.5 mm	37.5	50.0
25.0 mm	25.0	37.5
19.0 mm	19.0	25.0
12.5 mm	12.5	19.0
9.5 mm	9.5	12.5

Introduction   A factor that must be taken into account when considering asphalt mixture behavior is the *volumetric proportions* of asphalt binder and aggregate components, or more simply, *asphalt mixture volumetrics.* The developers of Superpave felt that the volumetric properties of asphalt mixtures were so important that a volumetric mixture design protocol was developed. This chapter describes volumetric analysis of HMA, which plays a significant role in most mixture design procedures, including the Superpave system.

The volumetric properties of a compacted paving mixture [air voids (V_a), voids in the mineral aggregate (VMA), voids filled with asphalt (VFA), and effective asphalt content (P_{be})] provide some indication of the mixture's probable pavement service performance. It is necessary to understand the definitions and analytical procedures described in this chapter to be able to make informed decisions concerning the selection of the design asphalt mixture. The information here applies to both paving mixtures that have been compacted in the laboratory and to undisturbed samples that have been cut from a pavement in the field.

Definitions   Mineral aggregate is porous and can absorb water and asphalt to a variable degree. Furthermore, the ratio of water to asphalt absorption varies with each aggregate. The three methods of measuring aggregate specific gravity consider these variations. These methods are bulk, apparent, and effective specific gravity:

Bulk Specific Gravity, G_{sb} – the ratio of the mass in air of a unit volume of a permeable material (including both permeable and impermeable voids normal to the material) at a stated temperature to the mass in air of equal density of an equal volume of gas-free distilled water at a stated temperature. See Figure 4.1.

Apparent Specific Gravity, G_{sa} – the ratio of the mass in air of a unit volume of an impermeable material at a stated temperature to the mass in air of equal density of an equal volume of gas-free distilled water at a stated temperature. See Figure 4.1.

Effective Specific Gravity, G_{se} – the ratio of the mass in air of a unit volume of a permeable material (excluding voids permeable to asphalt) at a stated temperature to the mass in air of equal density of an equal volume of gas-free distilled water at a stated temperature. See Figure 4.1.

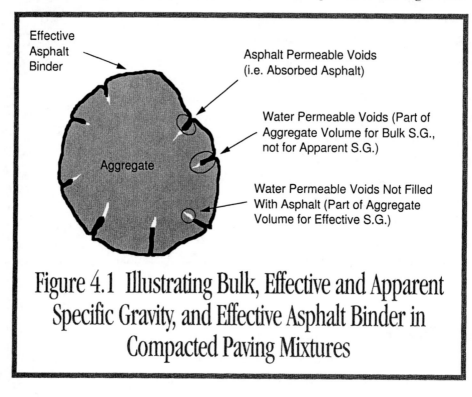

Effective Asphalt Binder

Asphalt Permeable Voids (i.e. Absorbed Asphalt)

Water Permeable Voids (Part of Aggregate Volume for Bulk S.G., not for Apparent S.G.)

Aggregate

Water Permeable Voids Not Filled With Asphalt (Part of Aggregate Volume for Effective S.G.)

Figure 4.1 Illustrating Bulk, Effective and Apparent Specific Gravity, and Effective Asphalt Binder in Compacted Paving Mixtures

The definitions for voids in the mineral aggregate (VMA), effective asphalt content (P_{be}), air voids (V_a), and voids filled with asphalt (VFA) are:

Voids in the Mineral Aggregate, VMA – the volume of intergranular void space between the aggregate particles of a compacted paving mixture that includes the air voids and the effective asphalt content, expressed as a percent of the total volume of the sample. See Figure 4.2.

Effective Asphalt Content, P_{be} – the total asphalt content of a paving mixture minus the portion of asphalt absorbed into the aggregate particles.

Air Voids, V_a – the total volume of the small pockets of air between the coated aggregate particles throughout a compacted paving mixture, expressed as percent of the bulk volume of the compacted paving mixture. See Figure 4.2.

Voids Filled with Asphalt, VFA – the percentage portion of the volume of intergranular void space between the aggregate particles that is occupied by the effective asphalt. It is expressed as the ratio of (VMA - V_a) to VMA. See Figure 4.2.

The Superpave mix design procedures require the calculation of VMA values for compacted paving mixtures in terms of the aggregate's bulk specific gravity, G_{sb}. Use of other aggregate specific gravities to compute

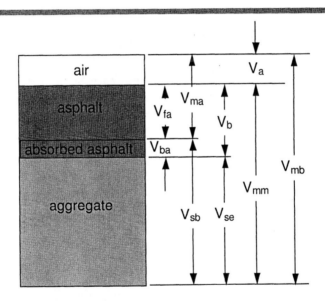

V_{ma} = Volume of voids in mineral aggregate
V_{mb} = Bulk volume of compacted mix
V_{mm} = Voidless volume of paving mix
V_{fa} = Volume of voids filled with asphalt
V_a = Volume of air voids
V_b = Volume of asphalt
V_{ba} = Volume of absorbed asphalt
V_{sb} = Volume of mineral aggregate (by bulk specific gravity)
V_{se} = Volume of mineral aggregate (by effective specific gravity)

Figure 4.2 Component Diagram of Compacted HMA Specimen

VMA means that the VMA criteria no longer apply. The effective specific gravity, G_{se}, should be the basis for calculating the air voids in a compacted asphalt paving mixture.

Voids in the mineral aggregate (VMA) and air voids (V_a) are expressed as percent by volume of the paving mixture. Voids filled with asphalt (VFA) is the percentage of VMA filled by the effective asphalt. Depending on how asphalt content is specified, the effective asphalt content may be expressed as either percent, by mass, of the total mass of the paving mixture, or as percent, by mass, of the aggregate in the paving mixture.

Because air voids, VMA and VFA are volume quantities, and therefore cannot be weighed, a paving mixture must first be designed or analyzed on a volume basis. For design purposes, this volume approach can easily be changed over to a mass basis to provide a job-mix formula.

Analyzing a Compacted Paving Mixture

The measurements and calculations needed for a voids analysis are:
(a) Measure the bulk specific gravity of the coarse aggregate (AASHTO T85 or ASTM C127) and of the fine aggregate (AASHTO T84 or ASTM C128).
(b) Measure the specific gravity of the asphalt cement (AASHTO T228 or ASTM D70)   and of the mineral filler (AASHTO T100 or ASTM D854).
(c) Calculate the bulk specific gravity of the aggregate combination in the paving mixture.
(d) Measure the maximum specific gravity of the loose paving mixture (ASTM D2041 or AASHTO T209).
(e) Measure the bulk specific gravity of the compacted paving mixture (ASTM D1188 / D2726 or AASHTO T166).
(f) Calculate the effective specific gravity of the aggregate.
(g) Calculate the maximum specific gravity at other asphalt contents.
(h) Calculate the asphalt absorption of the aggregate.
(i) Calculate the effective asphalt content of the paving mixture.
(j) Calculate the percent voids in the mineral aggregate in the compacted paving mixture.

(k) Calculate the percent air voids in the compacted paving mixture.

(l) Calculate the percent voids filled with asphalt in the compacted paving mixture.

Equations for these calculations are found below.

Table 4.1 illustrates example data for a sample of paving mixture. These design data are used for the sample calculations in the remainder of this chapter.

Table 4.1 Basic Data for Sample of Paving Mixture

Mixture Components

Material	Specific Gravity		Test Methods		Mix Composition	
		Bulk	AASHTO	ASTM	Percent by by Mass of Total Mix	Percent by by Mass of Total Aggr.
Asphalt Cement	1.030 (G_b)		T228	D70	5.3 (P_b)	5.6 (Pb)
Coarse Aggregate		2.716 (G_1)	T85	C127	47.4 (P_1)	50.0 (P_1)
Fine Aggregate		2.689 (G_2)	T84	C128	47.3 (P_2)	50.0 (P_2)
Mineral Filler	--		T100	D854	--	--

Paving Mixture
Bulk specific gravity of compacted paving mixture specimen, G_{mb}=2.442
Maximum specific gravity of paving mixture specimen, G_{mm}=2.535

Bulk Specific Gravity of Aggregate. When the total mass of aggregate consists of separate fractions of coarse aggregate, fine aggregate, and mineral filler, all having different measured specific gravities, the bulk specific gravity for the total aggregate is calculated using:

$$G_{sb} = \frac{P_1 + P_2 + ... + P_N}{\dfrac{P_1}{G_1} + \dfrac{P_2}{G_2} + ... + \dfrac{P_N}{G_N}}$$

where, G_{sb} = bulk specific gravity for the total aggregate

P_1, P_2, P_N = individual percentages by mass of aggregate

G_1, G_2, G_N = individual (e.g. coarse, fine) bulk specific gravity of aggregate

The bulk specific gravity of mineral filler is difficult to determine accurately. However, if the apparent specific gravity of the filler is substituted, the error is usually negligible.

Using the data in Table 4.1:

$$G_{sb} = \frac{50.0 + 50.0}{\dfrac{50.0}{2.716} + \dfrac{50.0}{2.689}} = \frac{100}{18.41 + 18.59} = 2.703$$

Effective Specific Gravity of Aggregate. When based on the maximum specific gravity of a paving mixture, G_{mm}, the effective specific gravity of the aggregate, G_{se}, includes all void spaces in the aggregate particles except those that absorb asphalt. G_{se} is determined using:

$$G_{se} = \frac{P_{mm} - P_b}{\dfrac{P_{mm}}{G_{mm}} - \dfrac{P_b}{G_b}}$$

Where, G_{se} = effective specific gravity of aggregate

G_{mm} = maximum specific gravity (ASTM D2041 / AASHTO T209) of paving mixture (no air voids)

P_{mm} = percent by mass of total loose mixture = 100

P_b = asphalt content at which ASTM D2041 / AASHTO T209 test was performed, percent by total mass of mixture

G_b = specific gravity of asphalt

Using the data in Table 4.1:

$$G_{se} = \frac{100 - 5.3}{\dfrac{100}{2.535} - \dfrac{5.3}{1.030}} = \frac{94.7}{39.45 - 5.15} = 2.761$$

> NOTE: The volume of asphalt binder absorbed by an aggregate is almost invariably less than the volume of water absorbed. Consequently, the value for the effective specific gravity of an aggregate should be between its bulk and apparent specific gravities. When the effective specific gravity falls outside these limits, its value must be assumed to be incorrect. The calculations, the maximum specific gravity of the total mix by ASTM D2041 / AASHTO T209, and the composition of the mix in terms of aggregate and total asphalt content should then be rechecked to find the source of the error.

Maximum Specific Gravity of Mixtures With Different

Asphalt Content. In designing a paving mixture with a given aggregate, the maximum specific gravity, Gmm, at each asphalt content is needed to calculate the percentage of air voids for each asphalt content. While the maximum specific gravity can be determined for each asphalt content by ASTM D2041 / AASHTO T209, the precision of the test is best when the mixture is close to the design asphalt content. In addition, it is preferable to measure the maximum specific gravity in duplicate or triplicate. Therefore, it is recommended that Gmm testing be performed on at least two mixture specimens at the estimated optimum asphalt content. The average of the measured Gmm values can be used to calculate Gmm at the other asphalt contents.

After calculating the effective specific gravity of the aggregate, from the measured maximum specific gravity, and averaging the Gse results, the maximum specific gravity for any other asphalt content can be obtained using the equation shown below. The equation assumes the effective specific gravity of the aggregate is constant; and, this is valid since asphalt absorption does not vary appreciably with changes in asphalt content.

$$G_{mm} = \frac{P_{mm}}{\dfrac{P_s}{G_{se}} + \dfrac{P_b}{G_b}}$$

where, G_{mm} = maximum specific gravity of paving mixture (no air voids)
P_{mm} = percent by mass of total loose mixture = 100
P_s = aggregate content, percent by total mass of mixture
P_b = asphalt content, percent by total mass of mixture
G_{se} = effective specific gravity of aggregate

Using the specific gravity data from Table 4.1, the effective specific gravity, G_{se}, and an asphalt content, P_b, of 4.0 percent:

$$G_{mm} = \frac{100}{\dfrac{96.0}{2.761} + \dfrac{4.0}{1.030}} = \frac{100}{34.77 + 3.88} = 2.587$$

Asphalt Absorption. Absorption is expressed as a percentage by mass of aggregate rather than as a percentage by total mass of mixture. Asphalt absorption, P_{ba}, is determined using:

$$P_{ba} = 100 \times \frac{G_{se} - G_{sb}}{G_{sb}G_{se}} \times G_b$$

where, P_{ba} = absorbed asphalt, percent by mass of aggregate
        G_{se} = effective specific gravity of aggregate
        G_{sb} = bulk specific gravity of aggregate
        G_b = specific gravity of asphalt

Using the bulk and effective aggregate specific gravities determined earlier and the asphalt specific gravity from Table 4.1:

$$P_{ba} = 100 \times \frac{2.761 - 2.703}{2.703 \times 2.761} \times 1.030 = 100 \times \frac{0.058}{7.463} \times 1.030 = 0.8$$

Effective Asphalt Content of a Paving Mixture. The effective asphalt content, P_{be}, of a paving mixture is the total asphalt content minus the quantity of asphalt lost by absorption into the aggregate particles. It is the portion of the total asphalt content that remains as a coating on the outside of the aggregate particles, and it is the asphalt content that governs the performance of an asphalt mixture. The formula is:

$$P_{be} = P_b - \frac{P_{ba}}{100} \times P_s$$

where, P_{be} = effective asphalt content, percent by total mass of mixture
        P_b = asphalt content, percent by total mass of mixture
        P_{ba} = absorbed asphalt, percent by mass of aggregate
        P_s = aggregate content, percent by total mass of mixture

Using the data from Table 4.1:

$$P_{be} = 5.3 - \frac{0.8}{100} \times 94.7 = 4.5$$

Percent VMA in Compacted Paving Mixture. The voids in the mineral aggregate, VMA, are defined as the intergranular void space between the aggregate particles in a compacted paving mixture that includes the air voids and the effective asphalt content, expressed as a percent of the total volume. The VMA is calculated based on the bulk specific gravity of the aggregate and is expressed as a percentage of the bulk volume of the compacted paving mixture. Therefore, the VMA can be calculated by subtracting the volume of the aggregate determined by its bulk specific gravity from the bulk volume of the compacted paving mixture. The calculation is illustrated for each type of mixture percentage content. If the mix composition is determined as percent by mass of total mixture:

$$VMA = 100 - \frac{G_{mb} \times P_s}{G_{sb}}$$

where, VMA = voids in the mineral aggregate, percent of bulk volume
G_{sb} = bulk specific gravity of total aggregate
G_{mb} = bulk specific gravity of compacted mixture
P_s = aggregate content, percent by total mass of mixture

Using the data from Table 4-1:

$$VMA = 100 - \frac{2.442 \times 94.7}{2.703} = 100 - 85.6 = 14.4$$

Or, if the mix composition is determined as percent by mass of aggregate:

$$VMA = 100 - \frac{G_{mb}}{G_{sb}} \times \frac{100}{100 + P_b} \times 100$$

where, P_b = asphalt content, percent by mass of aggregate

Using the data from Table 4-1:

$$VMA = 100 - \frac{2.442}{2.703} \times \frac{100}{100 + 5.6} \times 100 = 100 - 85.6 = 14.4$$

Percent Air Voids in Compacted Mixture. The air voids, V_a, in the total compacted paving mixture consist of the small air spaces between the coated aggregate particles. The volume percentage of air voids in a compacted mixture can be determined using:

$$V_a = 100 \times \frac{G_{mm} - G_{mb}}{G_{mm}}$$

where, Va = air voids in compacted mixture, percent of total volume

G_{mm} = maximum specific gravity of paving mixture (as determined in article 4.07 or as determined directly for a paving mixture by ASTM D2041 / AASHTO T209)

G_{mb} = bulk specific gravity of compacted mixture

Using the data from Table 4.1:

$$V_a = 100 \times \frac{2.535 - 2.442}{2.535} = 3.7$$

Percent VFA in Compacted Mixture. The percentage of the voids in the mineral aggregate that are filled with asphalt, VFA, not including the absorbed asphalt, is determined using:

$$VFA = 100 \times \frac{VMA - V_a}{VMA}$$

where, VFA = voids filled with asphalt, percent of VMA
VMA = voids in the mineral aggregate, percent of bulk volume
Va = air voids in compacted mixture, percent of total volume

Using the data from Table 4.1:

$$VFA = 100 \times \frac{14.4 - 3.7}{14.4} = 74.3$$

Introduction

Superpave mix design procedures involve:
▲ Selecting asphalt and aggregate materials that meet their respective criteria.
▲ Developing several aggregate trial blends to meet Superpave gradation requirements.
▲ Blending asphalt with the trial blends and short-term oven aging the mixtures.
▲ Compacting specimens and analyzing the volumetrics of the trial blends.
▲ Selecting "the best" trial blend as the design aggregate structure; and compacting samples of the design aggregate structure at several asphalt contents to determine the design asphalt content (Appendix C provides an outline of the mix design procedure).

The standardization of the Superpave mixture specimen preparation procedures has been initiated by AASHTO. Test procedures are found in AASHTO TP4, *Standard Method for Preparing and Determining the Density of Hot Mix Asphalt (HMA) Specimens by Means of the SHRP Gyratory Compactor*, and AASHTO PP2, *Standard Practice for Mixture Conditioning of Hot Mix Asphalt (HMA)*.

The testing procedures presented here are paraphrased from the AASHTO standards. The primary device used in Superpave mix design is the Superpave gyratory compactor (SGC). The SGC is used to produce specimens for volumetric analysis, and it also records data to provide a measure of specimen density throughout the compaction procedure.

➤➤ Superpave Gyratory Compactor

SHRP researchers had several goals in developing a laboratory compaction method. Most importantly, they wanted to realistically compact mixture test specimens to densities achieved under actual pavement climate and loading conditions. The compaction device needed to be capable of accommodating large aggregates and be able to measure compactability, so that potential tender mix behavior and similar compaction problems could be identified. A high priority for SHRP researchers was a device portable enough for use in the quality control operations of a mixing facility. Since no existing compactor achieved all these goals, the Superpave gyratory compactor (SGC) was developed.

The basis for the SGC was a Texas gyratory compactor modified to use the compaction principles of a French gyratory compactor. The modified Texas gyratory accomplished the goals of realistic specimen densification and it was reasonably portable. Its 6-inch sample diameter (ultimately 150 mm on an SGC) could accommodate mixtures containing aggregate up to 50-mm maximum size (37.5-mm nominal maximum size). SHRP researchers modified the Texas gyratory compactor by lowering its angle and speed of gyration and adding real time, specimen height-recording capabilities.

The SGC consists of these components:
- ▲ Reaction frame, rotating base, and motor
- ▲ Loading system, loading ram, and pressure gauge
- ▲ Height measuring and recordation system
- ▲ Mold and base plate
- ▲ Specimen extruding device

Figure 5.1 shows a schematic of the SGC.

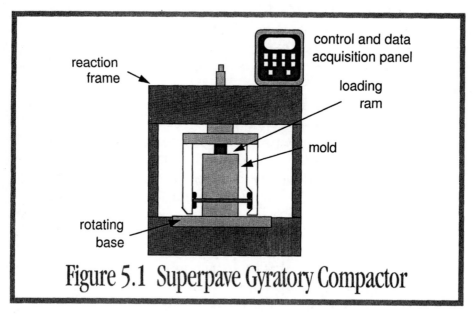

Figure 5.1 Superpave Gyratory Compactor

A loading mechanism presses against the reaction frame and applies a load to the loading ram to produce a 600-kPa compaction pressure on the specimen. A pressure gauge measures the ram loading to maintain constant pressure during compaction. The SGC mold (Figure 5.2) has an inside diameter of 150 mm and a base plate in the bottom of the mold provides confinement during compaction. The SGC base rotates at a constant rate of 30 revolutions per minute during compaction, with the mold positioned at a compaction angle of 1.25 degrees.

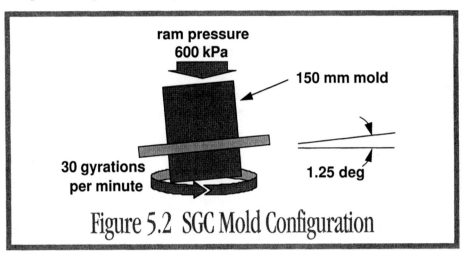

Figure 5.2 SGC Mold Configuration

Specimen height measurement is an important function of the SGC. Specimen density can be estimated during compaction by knowing the mass of material placed in the mold, the inside diameter of the mold, and the specimen height. Height is measured by recording the position of the ram throughout the test. Using these measurements, a specimen's compaction characteristic is developed (Figure 5.3).

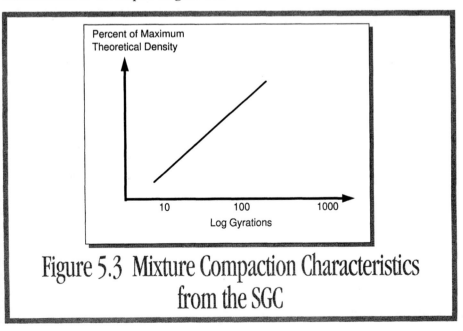

Figure 5.3 Mixture Compaction Characteristics from the SGC

Figure 5.3 illustrates how the density of the asphalt mixture increases with increasing gyrations. As with other mix design procedures, asphalt mixtures are designed at a specific level of compactive effort. In Superpave, this is a function of the design number of gyrations, N_{des}. N_{des} is used to vary the compactive effort of the design mixture and it is a function of traffic level. Traffic is represented by the design ESALs. The range of values for N_{des} is shown in Table 5.1.

Table 5.1 Superpave Gyratory Compactive Effort

Design ESALs (millions)	Compaction Parameters			Typical Roadway Applications
	N_{ini}	N_{des}	N_{max}	
< 0.3	6	50	75	Very light traffic (local / county roads; city streets where truck traffic is prohibited)
0.3 to < 3	7	75	115	Medium traffic (collector roads; most county roadways)
3 to < 30	8	100	160	Med. to high traffic (city streets; state routes; US highways; some rural interstates)
≥ 30	9	125	205	High traffic (most of the interstate system; climbing lanes; truck weighing stations.)

When specified by the agency and the top of the design layer is ≥ 100 mm from the pavement surface and the estimated design traffic level ≥ 0.3 million ESALs, decrease the estimated design traffic level by one, unless the mixture will be exposed to significant main line and construction traffic prior to being overlaid. If less than 25% of the layer is within 100 mm of the surface, the layer may be considered to below 100 mm for mixture design purposes.

When the design ESALs are between 3 to < 10 million ESALs the agency may, at their discretion, specify $N_{initial}$ at 7, N_{design} at 75, and N_{max} at 115, based on local experience.

Two other gyration levels are also of interest: the initial number of gyrations (N_{ini}), and maximum number of gyrations (N_{max}). Test specimens are compacted using N_{des} gyrations and an estimation of the compactability of the mixture is determined using N_{ini}. N_{max} is determined (using additional SGC specimens) after the mixture properties are established as a check to help guard against plastic failure caused by traffic in excess of the design level. N_{max} and N_{ini} are calculated from N_{des}, using the following relationships:

$$\text{Log } N_{max} = 1.10 \text{ Log } N_{des}$$
$$\text{Log } N_{ini} = 0.45 \text{ Log } N_{des}$$

The values of N_{ini}, N_{des} and N_{max} are shown for Superpave-defined traffic levels in Table 5.1.

SUPERPAVE MIX DESIGN

➤ ➤ Additional Test Equipment

Ancillary test equipment required in the preparation of Superpave asphalt mixtures includes:

1. Ovens, thermostatically controlled, for heating aggregates, asphalt, and equipment.
2. Mechanical Mixer: commercial bread dough mixer 10-liter (10-qt.) capacity or larger, equipped with metal mixing bowls and wire whips.
3. Flat-bottom metal pans for heating aggregates and aging mixtures.
4. Round metal pans, approximately 10-liter (10-qt.) capacity, for mixing asphalt and aggregate.
5. Scoops for batching aggregates.
6. Containers: gill-type tins, beakers, or pouring pots, for heating asphalt.
7. Thermometers: armored, glass, or dial-type with metal stem, 10°C to 235°C, for determining temperature of aggregates, asphalt and asphalt mixtures.
8. Balances: 10-kg capacity, sensitive to 1 g, for weighing aggregates and asphalt; 10-kg capacity, sensitive to 0.1 g, for weighing compacted specimens.
9. Large Mixing Spoon or small trowel.
10. Large spatula.
11. Welders gloves (or similar) for handling hot equipment.
12. Paint, markers, or crayons, for identifying test specimens.
13. Paper disks, 150 mm, for compaction.
14. Fans for cooling compacted specimens.
15. Computer/printer for data collection and recording.

Select Design Aggregate Structure

Prior to selecting a design aggregate structure, the individual asphalt and aggregate materials must be selected and approved. These procedures are discussed in Chapter 3.

To select the design aggregate structure, trial blends are established by mathematically combining the gradations of the individual aggregates into a single gradation. The blend gradation is then compared to the specification control requirements for the appropriate sieves. Appendix A contains the control points and restricted zone for the five Superpave mix gradations.

Trial blending consists of varying the stockpile percentages of each aggregate to obtain blend gradations meeting the gradation requirements for that particular mixture. There is no set number of trial blends that should initially be attempted. Three blends that cover a range of gradations are often sufficient for a starting point. Note that while Superpave recommends that gradations pass below the restricted zone, this is not a requirement. A trial gradation can plot above or even through the restricted zone. As an agency or contractor begins testing materials using the Superpave system, it would be beneficial to conduct analyses on many trial blends to determine the mixture behavior of the local materials.

Once the trial blends are selected, a preliminary evaluation of the blended aggregate properties is necessary. This includes the four consensus properties, the bulk and apparent specific gravity of the aggregate, and any source aggregate properties. These values can initially be mathematically estimated from the individual aggregate properties. Actual tests should be performed on the aggregate blend for final approval.

After the aggregate properties have been evaluated, the next step is to compact specimens and determine the volumetric properties of each trial blend. The trial asphalt binder content for each trial blend can be calculated by following the procedure detailed in AASHTO Provisional Standard, PP-28, Appendix XI.

Rather than calculate the trial asphalt binder contents, many designers prefer to estimate the trial values based on experience or other information. The following values are typical for aggregate blends having combined aggregate bulk specific gravity of approximately 2.65. Aggregate combinations having significantly higher G_{sb} may need less asphalt and those with a lower G_{sb} may need more asphalt.

Nominal Maximum Aggregate Size (mm)	Trial Asphalt Binder Content (%)
37.5	3.5
25.0	4.0
19.0	4.5
12.5	5.0
9.5	5.5

A minimum of two specimens for each trial blend is compacted using the Superpave gyratory compactor. Two samples are also prepared for determining the mixture's maximum theoretical specific gravity. An aggregate mass of 4700 grams is usually sufficient for the compacted specimens. An aggregate mass of 2000 grams is usually sufficient for the specimens used to determine maximum theoretical specific gravity (G_{mm}), although AASHTO T209 (ASTM D2041) should be consulted to determine the minimum sample size required for various mixtures.

Specimen Preparation and Compaction

This procedure outlines preparation of HMA test specimens using the Superpave gyratory compactor (SGC). It includes guidelines for mixing and compacting test specimens. Figure 5.5 illustrates the various steps in specimen preparation and compaction.

Preparation of Aggregates. Prepare a batching sheet showing the batch weights of each aggregate component and the asphalt binder. Weigh into a pan the proper weights of each aggregate component.

Three sample sizes are used depending on their final use. For compacted specimens that will be used in Superpave mix design, the specimen size is 150 mm (diameter) by 115 mm (height) and requires approximately 4700 g of aggregate. Samples to be used for determination of maximum theoretical specific gravity by AASHTO T209 / ASTM D2041 remain uncompacted and their sizes vary by aggregate size and range from 1000 to 2500 grams. Moisture damage testing using AASHTO T283 requires a specimen height of 95 mm and approximately 3700 g of aggregate.

Mixing and Compaction Temperatures. Determine the laboratory mixing and compaction temperatures using a plot of viscosity versus temperature (Figure 5.4). Select mixing and compaction temperatures corresponding with binder viscosity values of of 0.17 ± 0.02 Pa•s and 0.28 ± 0.03 Pa•s, respectively.

Note that these viscosity ranges are not valid for modified asphalt binders. The designer should consider the manufacturer's recommendations when establishing mixing and compaction temperatures for modified binders. Practically, the mixing temperature should not exceed 165°C and the compaction temperature should not be lower than 115°C.

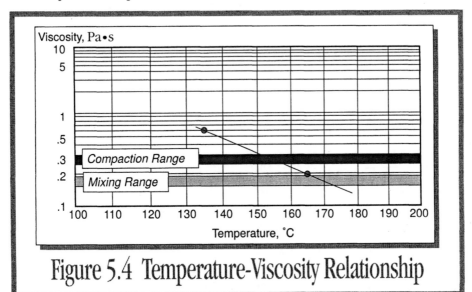

Figure 5.4 Temperature-Viscosity Relationship

Note: *The mixing and compaction temperatures described above are intended for laboratory mix design purposes only. Field mixing and compaction temperatures must be determined from trial applications and experience.*

Place the pan containing the aggregate in an oven set approximately 15°C higher than the mixing temperature. Two to four hours are required for the aggregate to reach the mixing temperature. While aggregate is heating, heat all mixing implements such as spatulas, mixing bowl, and other tools. Heat the asphalt binder to the desired mixing temperature. The time required for this step varies depending on the amount of asphalt and the heating method.

Preparation of Mixtures.

▲ Place the hot mixing bowl on a balance and zero the balance (Figure 5.5 A).
▲ Charge the mixing bowl with the heated aggregates and mix thoroughly.
▲ Form a crater in the blended aggregate and weigh the required asphalt into the mixture to achieve the desired batch weight.
▲ Remove the mixing bowl from the scale and mix the asphalt and aggregate using a mechanical mixer (Figure 5.5 B).
▲ Mix the specimen until the aggregate is thoroughly coated.
▲ Place the mix in a flat shallow pan at an even thickness ranging between 25 mm and 50 mm.
▲ Place the mix and pan in the conditioning oven for 2 hours ± 5 minutes at a temperature equal to the mixture's specified compaction temperature ± 3°C (Figure 5.5 C).
▲ *Short-term age the specimen for 2 hours.
▲ Repeat this procedure until the desired number of specimens are produced.
▲ Proper timing of the gyratory compaction steps can be achieved by spacing approximately 20 minutes between mixing each specimen.
▲ At the end of the short-term aging period, proceed to AASHTO T209 / ASTM D2041 if the mixture is to be used to determine maximum theoretical specific gravity. Otherwise, proceed with compaction.

*Experience has shown that aggregates having more than 2 percent (water) absorption should be aged for four hours to allow additional asphalt absorption to occur. This adjustment should more closely align laboratory values with volumetric calculations and measurements made on plant-produced mixture determined in production. For evaluation of properties of specimens compacted from mixing-plant produced asphalt mixtures containing absorptive aggregates, consideration should be given to adding a short-term aging (STOA) period. STOA periods of up to two hours may be necessary to allow asphalt binder absorption to occur.

Figure 5.5 A

Figure 5.5 B

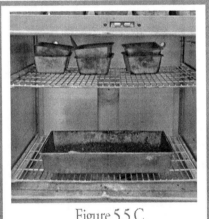

Figure 5.5 C

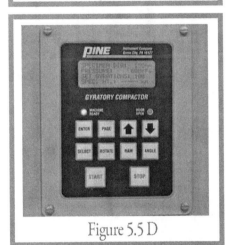

Figure 5.5 D

Figure 5.5 E

Figure 5.5 F

Figure 5.5 G

Figure 5.5 H

Figure 5.5 I

Figure 5.5 J

Figure 5.5 K

Figure 5.5 L

Compaction of Volumetric Specimens. Prepare the compactor while mixture specimens are short-term aging (Figure 5.5 D). This includes verifying that the compaction pressure, the compaction angle and speed of gyration are set to their proper values, and setting the desired number of gyrations, N_{des}. Also ensure that the data acquisition device is functioning.

Approximately 45-60 minutes before compaction of the first specimen, place the compaction molds and base/top plates in an oven set at the compaction temperature. Remove the mold and base plate from the oven and place the base plate in the mold and a paper disk on top of the base plate (Figure 5.5 E).

Place the short-term aged mixture in the mold (Figure 5.5 F), level the mixture, and place a paper disk on top of the leveled mixture. Place the mold containing the specimen into the compactor (Figures 5.5 G and 5.5 H). Center the mold under the loading ram and start the system so that the ram extends down into the mold cylinder and contacts the specimen. The ram will stop when the pressure reaches 600 kPa.

Apply the 1.25° angle of gyration and start the gyratory compaction (Figure 5.5 I). Compaction will proceed until N_{des} has been completed. During compaction, the ram loading system will maintain a constant pressure of 600 kPa. Specimen height is continually monitored during compaction and a height measurement is recorded after each revolution.

The compactor will cease compacting after reaching N_{des}, and the angle of gyration will be released and the loading ram raised. Remove the mold containing the compacted specimen from the compactor and slowly extrude the specimen from the mold (Figure 5.5 J). A 5-minute cooling period will facilitate specimen removal without undue distortion.

Remove the paper disk from the top and bottom of the specimen (Figure 5.4 K) and allow the specimen to cool undisturbed. Place the mold and base plate back in the oven to reach compaction temperature for the next specimen. Additional molds will avoid the delay caused by this step. Repeat the compaction procedure for each specimen. Identify each specimen with a suitable marker (Figure 5.5 L).

Data Analysis and Presentation

Superpave gyratory compaction data is analyzed by computing the estimated bulk specific gravity, corrected bulk specific gravity, and corrected percentage of maximum theoretical specific gravity for each desired gyration. During compaction, the height is measured and recorded after each gyration. G_{mb} of the compacted specimen and G_{mm} of the loose mixture are measured and an estimate of G_{mb}, at any value of gyration, is made by dividing the mass of the mixture by the volume of the compaction mold:

$$G_{mb}\,(estimated\,) = \frac{W_m / V_{mx}}{\gamma_w}$$

where, G_{mb}(estimated) = estimated bulk specific gravity of specimen during compaction

W_m = mass of specimen, grams

γ_w = density of water = 1 g/cm^3

V_{mx} = volume of compaction mold (cm^3), calculated using the equation:

where, d = diameter of mold (150 mm), and

h_x = height of specimen in mold during compaction (mm)

π = 3.1416

This calculation assumes that the specimen is a smooth-sided cylinder, which it is not. Surface irregularities cause the volume of the specimen to be slightly less than the volume of a smooth-sided cylinder. Therefore, the final estimated G_{mb} at N_{des} is different than the measured G_{mb} at N_{des}. Therefore, the estimated G_{mb} is corrected by a ratio of the measured to estimated bulk specific gravity:

where, C = correction factor

$$C = \frac{G_{mb}\,(measured\,)}{G_{mb}\,(estimated\,)}$$

G_{mb}(measured) = measured bulk specific gravity after N_{des}

G_{mb}(estimated) = estimated bulk specific gravity at N_{des}

The estimated G_{mb} at any other gyration level is then determined using:

$$G_{mb}\,(corrected\,) = C \times G_{mb}\,(estimated\,)$$

where, G_{mb} (corrected) = corrected bulk specific gravity of the specimen at any gyration

C = correction factor, and

G_{mb} (estimated) = estimated bulk specific gravity at any gyration.

Percent G_{mm} at any gyration level is then calculated as the ratio of G_{mb} (corrected) to G_{mm} (measured), and the average percent G_{mm} values for the two companion specimens are calculated.

Using the N_{max}, N_{des}, and N_{ini} gyrations levels previously determined from design traffic level, Superpave volumetric mix design criteria (VMA, VFA, and dust ratio) are established on a four percent air void content at N_{des}. Superpave mix design also specifies criteria for the mixture density at N_{ini}, Nd_{es}, and N_{max}.

The percent air voids at Ndes is determined from the equation:

$$V_a = 100 - \%G_{mm} @ N_{des}$$

where, Va $\quad = \quad$ air voids @ N_{des}, percent of total volume

$\%G_{mm} @ N_{des} \quad = \quad$ maximum theoretical specific gravity @ N_{des}, percent

The percent voids in the mineral aggregate is calculated using:

$$\%VMA = 100 - (\frac{\%G_{mm}@\,N_{des} \times G_{mm} \times P_s}{G_{sb}})$$

where, VMA $\quad = \quad$ voids in the mineral aggregate, percent of bulk volume

$\%G_{mm} @ N_{des} \quad = \quad$ maximum theoretical specific gravity @ N_{des}, percent

$G_{mm} \quad = \quad$ maximum theoretical specific gravity

$G_{sb} \quad = \quad$ bulk specific gravity of total aggregate

$P_s \quad = \quad$ aggregate content, cm^3/cm^3, by total mass of mixture

If the percentage of air voids is equal to four percent, then this data is compared to the volumetric criteria and an analysis of this blend completed. However, if the air void content at N_{des} varies from four percent (and this will typically be the case), an estimated design asphalt content to achieve 4 percent air voids at N_{des} is determined; and, the estimated design properties at this estimated design asphalt content are calculated.

The estimated asphalt content for: N_{des} = four percent air voids is calculated using this equation:

$$P_{b\ estimated} = P_{bi} - (0.4 \times (4-V_a))$$

where, $P_{b,\ estimated}$ = estimated asphalt content, percent by mass of mixture

P_{bi} = initial (trial) asphalt content, percent by mass mixture

V_a = percent air voids at N_{des} (trial)

The volumetrics (VMA and VFA) at N_{des} and mixture density at N_{ini} are then estimated at this asphalt binder content using the equations that follow.

For VMA:

$$\%VMA_{estimated} = \%VMA_{initial} + C x (4 - V_a)$$

where, $\%VMA_{initial}$ = %VMA from trial asphalt binder content

C = constant = *0.1 if Va is less than 4.0 percent*

= *0.2 if Va is greater than 4.0 percent*

Specified minimum values for VMA at the design air void content of four percent are a function of nominal maximum aggregate size and are given in Table 5.2.

For VFA:

$$\%VFA_{estimated} = 100 \times \frac{(\%VMA_{estimated} - 4.0)}{\%VMA_{estimated}}$$

The acceptable range of design VFA at four percent air voids as a function of traffic level is shown in Table 5.2.

For $\%G_{mm}$ at N_{ini}:

$$\%G_{mm\ estimated} @ N_{ini} = \% G_{mm\ trial} @ N_{ini} - (4.0 - V_a)$$

The maximum allowable mixture density at N_{ini} for the various traffic levels is shown in Table 5.2.

Table 5.2 Superpave Volumetric Mixture Design Requirements

Design ESALs (million)	Required Density (% of Theoretical Maximum Specify Gravity)			Voids-in-the Mineral Aggregate (Percent), minimum					Voids Filled With Asphalt (Percent)	Dust-to-Binder Ratio
				Nominal Maximum Aggregate Size, mm						
	$N_{initial}$	N_{design}	N_{max}	37.5	25.0	19.0	12.5	9.5		
< 0.3	≤ 91.5								70 - 80	
0.3 to < 3	≤ 90.5								65 - 78	
3 to < 10		96.0	≤ 98.0	11.0	12.0	13.0	14.0	15.0	65 - 75	0.6 - 1.2
10 to < 30	≤ 89.0									
≥ 30										

Design ESALs are the anticipated project traffic level expected on the design lane over a 20-year period. Regardless of the actual design life of the roadway, determine the design ESALs for 20 years, and choose the appropriate N_{design} level.

For 9.5-mm nominal maximum size mixtures, the specified VFA range shall be 73% to 76% for design traffic levels ≥ 3 million ESALs.

For 25.0-mm nominal maximum size mixtures, the specified lower limit of the VFA shall be 67% for design traffic levels < 0.3 million ESALs.

For 37.5-mm nominal maximum size mixtures, the specified lower limit of the VFA shall be 64% for all design traffic levels.

If the aggregate gradation passes beneath the boundaries of the aggregate restricted zone, consideration should be given to increasing the dust-to-binder ratio criteria from 0.6 - 1.2 to 0.8 - 1.6

Finally, there is a requirement for the dust proportion. It is calculated as the percent by mass of the material passing the 0.075-mm sieve (by wet sieve analysis) divided by the effective asphalt binder content (expressed as percent by mass of mix). The effective asphalt binder content is calculated using:

$$P_{be} = -(P_s \times G_b) \times (\frac{G_{se} - G_{sb}}{G_{se} \times G_{sb}}) + P_{b,\,estimated}$$

where, P_{be} = effective asphalt content, percent by total mass of mixture

P_s = aggregate content, percent by total mass of mixture

G_b = specific gravity of asphalt

G_{se} = effective specific gravity of aggregate

G_{sb} = bulk specific gravity of aggregate

P_b = asphalt content, percent by total mass of mixture

Dust Proportion is calculated using:

$$DP = \frac{P_{0.075}}{P_{be}}$$

where, $P_{0.075}$ = aggregate content passing the 0.075-mm sieve, percent by mass of aggregate

P_{be} = effective asphalt content, percent by total mass of mixture

An acceptable dust proportion ranges from 0.6 to 1.2. However, consideration may be given to increasing the ratio to 0.8 to 1.6 if the aggregate gradation passes below the restricted zone

After establishing all the estimated mixture properties, the designer can look at the trial blends and decide if one or more are acceptable, or if further trial blends need evaluation.

Design Asphalt Binder Content

Once the design aggregate structure is selected from the trial blends, specimens are compacted at varying asphalt binder contents. The mixture properties are then evaluated to determine a design asphalt binder content.

A minimum of two specimens are compacted at the trial blend's estimated asphalt content, at ± 0.5% of the estimated asphalt content, and at + 1.0% of the estimated asphalt content. The four asphalt contents are the minimum required for Superpave mix design.

A minimum of two specimens are also prepared for determination of maximum theoretical specific gravity at the estimated binder content. Specimens are prepared and tested in the same manner as the specimens from the "Select Design Aggregate Structure" section.

Mixture properties are evaluated for the selected blend at the different asphalt binder contents, by using the densification data at N_{ini} and N_{des}. The volumetric properties are calculated at N_{des} for each asphalt content. From these data points, the designer can generate graphs of air voids, VMA, and VFA versus asphalt content. The design asphalt binder content is established at 4.0 percent air voids. All other mixture properties are checked at the design asphalt binder content to verify that they meet criteria.

After verifying that all other mixture properties meet criteria at the design asphalt binder content, two additional SGC specimens are compacted to N_{max} (from table 5.1) to ensure that N_{max} does not exceed 98.0% G_{mm}.

Moisture Sensitivity

The final step in the Superpave mix design process is to evaluate the moisture sensitivity of the design mixture. This step is accomplished by performing the AASHTO T283 test, *Resistance of Compacted Bituminous Mixtures to Moisture Induced Damage* on the design aggregate blend at the design asphalt binder content. Specimens for this test are compacted to approximately 7 percent air voids. One subset, consisting of three specimens, is considered the control set. The other subset of three specimens is conditioned.

The conditioned specimens are subjected to partial vacuum saturation followed by an optional freeze cycle, followed by a 24-hour thaw cycle at 60°C. All specimens are tested to determine their indirect tensile strengths. The moisture sensitivity is determined as a ratio of the average tensile strengths of the conditioned subset divided by the average tensile strengths of the control subset. The Superpave criterion for tensile strength ratio is 80 percent, minimum.

Introduction This chapter presents a full Superpave volumetric mix design example.

Volumetric mix design plays a central role in Superpave mixture design. The best way of illustrating its steps is by means of an example. This section provides the Superpave mixture design test results for a project that was constructed in 1992 by the Wisconsin Department of Transportation on Interstate 43 in Milwaukee. The information presented follows the logical progression of testing and data analysis involved in a Superpave mixture design and encompasses the concepts outlined in previous sections. There are four major steps in the testing and analysis process:

1. selection of materials (aggregates, binders, modifiers, etc.),
2. selection of a design aggregate structure,
3. selection of a design asphalt binder content,
4. evaluation of moisture sensitivity of the design mixture.

Selection of materials begins with determining the traffic and environmental factors for the paving project. From that information, the performance grade of asphalt binder required for the project is selected. Aggregate requirements are determined based on traffic level and layer depth. Materials are selected based on their ability to meet or exceed the established criteria.

Selection of the design aggregate structure is accomplished by comparing the properties of a series of trial mixtures. This step consists of blending available aggregate stockpiles at different percentages to arrive at aggregate gradations that meet Superpave requirements. Three trial blends are normally employed for this purpose. A trial blend is considered acceptable if it possesses suitable volumetric properties (based on traffic and environmental conditions) at an appropriate design binder content. Once selected, the trial blend becomes the design aggregate structure.

Selection of a design asphalt binder content consists of varying the amount of asphalt binder with the design aggregate structure to obtain acceptable volumetric and compaction properties when compared to the mixture criteria, which are based on traffic and environmental conditions. This step is a verification of the results obtained from the previous step. This step also allows the designer to observe the sensitivity of volumetric and compaction properties of the design aggregate structure to asphalt content. The design aggregate structure at the design asphalt binder content becomes the job-mix formula.

Evaluation of moisture sensitivity consists of testing the designed mixture by AASHTO T283 to determine if the mix will be susceptible to moisture damage.

Material Selection   For the IH-43 project, design ESALs are determined to be 18 million in the design lane. This places the design in the traffic category from 3 to 30 million ESALs. Traffic level is used to determine design requirements such as number of design gyrations for compaction, aggregate physical property requirements, and mixture volumetric requirements.

The mixture in this example is an intermediate course mixture. It will have a nominal maximum particle size of 19.0 mm. It will be placed at a depth less than 100 mm from the surface of the pavement.

➤➤ Binder Selection

Environmental conditions are determined from weather station data stored in the Superpave weather database. The data can be retrieved from the report Weather Database for the Superpave Mix Design System, SHRP-A-648A, or from the LTPPBIND software released by the Long-Term Pavement Performance (LTPP) Division of the FHWA. The project near Milwaukee has two weather stations as depicted in Table 6.1:

Table 6.1 Project Environmental Conditions and Binder Grades				
Weather Station	Min. Pvmt. Temp. (°C)	Max. Pvmt. Temp. (°C)	Binder Grade	Design High Air Temp. (°C)
Low Reliability (50%)				
Milwaukee Mt. Mary	-26	52	PG 52-28	32
Milwaukee WSO AP	-25	51	PG 52-28	31
Paving Location	**-26**	**52**	**PG 52-28**	**32**
High Reliability (98%)				
Milwaukee Mt. Mary	-32	55	PG 58-34	36
Milwaukee WSO AP	-33	54	PG 58-34	34
Paving Location	**-32**	**55**	**PG 58-34**	**35**

Low and high reliability level binder grades are shown. Reliability is the percent probability that the actual temperature will not exceed the design pavement temperatures listed in the binder grade. In this example, the designer chooses high reliability for all conditions. Thus, a PG 58-34 binder is needed. The average Design High Air Temperature is 35°C.

Having determined the need for a PG 58-34 binder, the binder is selected and tested for specification compliance. Binder test results are summarized in Table 6.2:

Table 6.2 Binder Test Results

Test	Parameter	Test Result	Criteria
	Original Binder		
Flash Point	n/a	304°C	230°C minimum
Rotational Viscosity	135°C	0.500 Pa·s	3 Pa·s maximum
Rotational Viscosity	165°C	0.075 Pa·s	n/a
Dynamic Shear Rheometer	G*/sin δ @ 58°C	1.42 kPa	1.00 kPa minimum
	RTFO-aged Binder		
Mass Loss	n/a	0.14%	1.00% maximum
Dynamic Shear Rheometer	G*/sin δ @ 58°C	2.41 kPa	2.20 kPa minimum
	PAV-aged Binder		
Dynamic Shear Rheometer	G*/sin δ @ 16°C	1543 kPa	5000 kPa maximum
Bending Beam Rheometer	Stiffness @ -24°C	172.0 MPa	300.0 MPa maximum
Bending Beam Rheometer	m-value @ -24°C	0.321	0.300 minimum

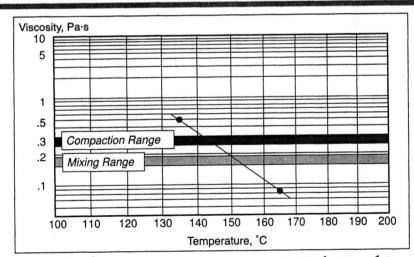

Figure 6.1 Temperature-Viscosity Relationship for PG 58-34 Binder

Comparing the test results to specifications, the designer verifies that the asphalt binder meets the requirements of a PG 58-34 grade. Specification testing requires only that rotational viscosity be performed at 135°C. Additional testing was performed at 165°C to establish laboratory mixing and compaction temperatures. The illustration of the temperature-viscosity relationship for this binder shows that the mixing temperature range is selected between 148°C and 153°C. The compaction temperature range is selected between 142°C and 146°C.

➤ ➤ Aggregate Selection

Next, the designer selects the aggregates to use in the mixture. For this example, there are five stockpiles of materials consisting of three coarse materials and two fine materials. It is assumed that the mixing facility will have at least five cold feed bins. If fewer cold feed bins are available, fewer stockpiles will be used. The materials are split into representative samples, and a washed sieve analysis is performed for each aggregate.

The bulk and apparent specific gravity are determined for each aggregate. These specific gravity values are used in VMA calculations and may be used if trial binder contents are calculated. Aggregate specific gravity values for this example listed in Table 6.3

Table 6.3 Aggregate Specific Gravity		
Aggregate	Bulk Sp. Gr.	Apparent Sp. Gr.
#1 Stone	2.703	2.785
12.5mm Chip	2.689	2.776
9.5 mm Chip	2.723	2.797
Manuf. Sand	2.694	2.744
Screen Sand	2.679	2.731

In addition to sieve analysis and specific gravity determination, Superpave requires that consensus aggregate tests be performed to assure that the aggregates selected for the mix design are acceptable. The four tests required are: coarse aggregate angularity, fine aggregate angularity, thin and elongated particles, and clay content. In addition, the specifying agency can select any other aggregate tests deemed important. These tests can include items such as soundness, toughness, and deleterious materials among others.

Superpave consensus aggregate criteria are applied to combined aggregate blends rather than individual aggregate components. However, some designers find it useful to perform the aggregate tests on the individual aggregate components. This step allows the designer to use the test results in narrowing the acceptable range of blend percentages for the aggregates. It also allows for greater flexibility if multiple trial blends are attempted. The test results from the components can be used to estimate the results for a given combination of materials. However, when the design aggregate structure is selected, the properties of the selected blend will need to be verified by testing. The drawback to this procedure is that it takes more time to perform this additional testing. For this example, the aggregate properties are measured for each stockpile as well as for the aggregate trial blends.

Coarse Aggregate Angularity. This test is performed on the coarse aggregates. The coarse aggregate particles are defined as those larger than 4.75 mm. Table 6.4 lists the results for coarse aggregate angularity for this example.

Table 6.4 Coarse Aggregate Angularity Test Results, %				
Aggregate	1+ Fractured Faces	Criteria	2+ Fractured Faces	Criteria
#1 Stone	92		88	
12.5 mm Chip	97	95 (min)	94	90 (min)
9.5 mm Chip	99		95	

Table 6.4 also shows the criteria for fractured faces based on traffic (18 million ESALs) and depth from the surface (< 100 mm). The criteria change as the traffic level and layer position (relative to the surface) change. Note that the #1 Stone does not meet either of the fractured faces criteria. However, this material can be used as long as the selected *blend* of aggregates meets the design criteria.

Fine Aggregate Angularity. This test is performed on the fine aggregate. The fine aggregate particles are defined as those smaller than 2.36 mm. Table 6.5 lists the results for fine aggregate angularity.

Table 6.5 Fine Aggregate Angularity		
Aggregate	% Air Voids (Loose)	Criteria %
Manufactured Sand	52	45 (min)
Screen Sand	40	

Table 6.5 also indicates the criteria for fine aggregate angularity based on traffic and depth from the surface. Even though the screen sand tested below the minimum criteria, it can be used as long as the selected *blend* of aggregates meet the requirement.

Flat, Elongated Particles. This test is performed on the coarse aggregates. The coarse aggregate particles are defined as those larger than 4.75 mm. Table 6.6 contains the results for flat and elongated particles.

Table 6.6 Flat, Elongated Particles		
Aggregate	% Flat / Elongated	Criteria, %
#1 Stone	0	
12.5 mm Chip	0	10 (max)
9.5 mm Chip	0	

Table 6.6 shows the criteria for percentage of flat and elongated particles, which is based on traffic only. The criteria changes as the traffic level changes. In this case, the aggregates readily satisfy the requirement.

Clay Content (Sand Equivalent). This test is performed on the fine aggregate. The fine aggregate particles are defined as those smaller than 4.75 mm. Test results for sand equivalent are contained in Table 6.7.

Table 6.7 Clay Content (Sand Equivalent)		
Aggregate	Sand Equivalent, %	Criteria, %
Manufactured Sand	47	45 (min)
Screen Sand	70	

Table 6.7 also indicates the criteria for sand equivalent which is based on the traffic volume. The sand equivalent results for both fine aggregates exceed the minimum requirement, so there is a reasonable expectation that the blend, when tested, will also meet the clay content requirement.

Once all of the aggregate criteria are met, the materials selection process is complete. The next step is to select the design aggregate structure.

Select Design Aggregate Structure

To select the design aggregate structure, the designer establishes trial blends by mathematically combining the gradations of the individual materials into a single gradation. The blend gradation is then compared to the specification requirements for the appropriate sieves. Gradation control is based on four control sieves: the maximum sieve, the nominal maximum sieve, the 2.36-mm sieve, and the 0.075-mm sieve.

The nominal maximum sieve is one sieve size larger than the first sieve to retain more than ten percent of combined aggregate. The maximum sieve size is one sieve size greater than the nominal maximum sieve. The restricted zone is an area on either side of the maximum density line. For a 19.0-mm nominal mixture, it starts at the 2.36-mm sieve and extends to the 0.300-mm sieve. Any proposed trial blend gradation has to pass between the control points established for the four sieves. In addition, it is recommended to be outside of the restricted zone. Some specifying agencies are allowing gradations to pass through the restricted zone if there is a history of successful performance or supporting test results. Table 6.8 illustrates the gradation control criteria for a 19.0-mm nominal mixture.

Table 6.8 Gradation Criteria for 19.0 mm Nominal Mixture			
Gradation Control Item	Sieve Size, mm	Minimum, %	Maximum, %
Control Points	25.0	100.0	100.0
	19.0	90.0	100.0
	12.5		90.0
	2.36	23.0	49.0
	0.075	2.0	8.0
Restricted Zone	2.36	34.6	34.6
	1.18	22.3	28.3
	0.600	16.7	20.7
	0.300	13.7	13.7

Trial blending consists of varying the cold-feed percentages of the available aggregates to obtain blended gradations meeting the requirements for that particular mixture. Any number of trial blends may be evaluated, but a minimum of three trial blends is required. For this example, three trial blends are used: an intermediate blend (Blend 1), a coarse blend (Blend 2), and a fine blend (Blend 3). The intermediate blend is combined to produce a gradation that is not close to any of the control point limits. The coarse blend is combined to produce a gradation that is near the minimum allowable percent passing the nominal maximum sieve, the 2.36-mm sieve, and the 0.075-mm sieve. The fine blend is combined to produce a gradation that is close to the maximum percent passing the nominal maximum size and is just below the restricted zone.

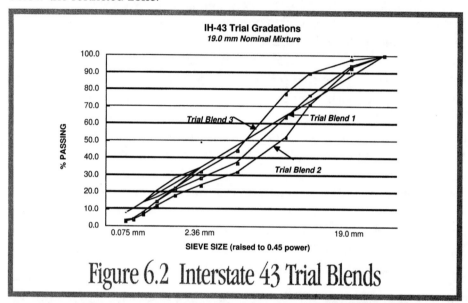

Figure 6.2 Interstate 43 Trial Blends

Table 6.9 IH-43 Trial Gradations,%

	#1 Stone	12.5 mm Chip	9.5 mm Chip	Mfg. Sand	Scr. Sand		
Blend 1	25.0	15.0	22.0	18.0	20.0		
Blend 2	30.0	25.0	13.0	17.0	15.0		
Blend 3	10.0	15.0	30.0	31.0	14.0		

Sieve							Blend 1 Grad.	Blend 2 Grad.	Blend 3 Grad.
25.0 mm	100.0	100.0	100.0	100.0	100.0		100.0	100.0	100.0
19.0 mm	76.1	100.0	100.0	100.0	100.0		94.0	92.8	97.6
12.5 mm	14.3	87.1	100.0	100.0	100.0		76.6	71.1	89.5
9.5 mm	3.8	26.0	94.9	100.0	99.8		63.7	51.9	77.7
4.75 mm	2.1	3.1	4.8	95.5	89.5		31.7	31.7	44.3
2.36 mm	1.9	2.6	3.0	63.5	76.7		28.3	23.9	31.9
1.18 mm	1.9	2.4	2.8	38.6	63.5		21.1	17.6	22.2
.600 mm	1.8	2.3	2.6	21.9	45.6		14.4	12.0	14.5
.300 mm	1.8	2.2	2.5	11.0	23.1		7.9	6.8	7.9
.150 mm	1.7	2.1	2.4	5.7	8.4		4.0	3.6	4.1
.075 mm	1.6	1.9	2.2	5.7	4.7		3.1	2.9	3.5

All three of the trial blends are shown graphically in Figure 6.2. Note that while all three trial blends pass below the restricted zone, this is not a requirement. Table 6.9 shows the gradations of the three trial blends.

Once the trial blends are selected, a preliminary determination of the blended aggregate properties is necessary. This can be estimated mathematically from the aggregate properties.

Table 6.10 Estimated Aggregate Blend Properties

Property	Criteria,%	Trial Blend 1,%	Trial Blend 2,%	Trial Blend 3,%
Coarse Ang.	95/90 min.	96/92	95/92	97/93
Fine Ang.	45 min.	46	46	48
Flat/Elongated	10 max.	0	0	0
Sand Equivalent	45 min.	59	58	54
Combined G_{sb}	n/a	2.699	2.697	2.701
Combined G_{sa}	n/a	2.768	2.769	2.767

Values for coarse aggregate angularity are shown as percentage of one or more fractured faces followed by percentage of two or more fractured faces. Based on the estimates, all three trial blends are acceptable. When the design aggregate structure is selected, the blend aggregate properties will need to be verified by testing. The estimated aggregate blend properties are summarized in Table 6.10.

Select Trial Asphalt Binder Content

The next step is to evaluate the trial blends by compacting specimens and determining the volumetric properties of each trial blend. For each blend, a minimum of two specimens will be compacted using the SGC. The trial asphalt binder content can be estimated based on experience with similar materials. If there is no experience, the trial binder content can be determined for each trial blend by estimating the effective specific gravity of the blend and using the calculations shown below. The effective specific gravity (G_{se}) of the blend is estimated by:

$$G_{se} = G_{sb} + 0.8 \, (\, (G_{sa} - G_{sb})$$

The factor, 0.8, can be adjusted at the discretion of the designer. Absorptive aggregates may require values closer to 0.6 or 0.5. The blend calculations are shown below:

Blend 1: $G_{se} = 2.699 + 0.8((2.768 - 2.699) = 2.754$
Blend 2: $G_{se} = 2.697 + 0.8((2.769 - 2.697) = 2.755$
Blend 3: $G_{se} = 2.701 + 0.8((2.767 - 2.701) = 2.754$

The volume of asphalt binder (V_{ba}) absorbed into the aggregate is estimated using this equation:

$$V_{ba} = \frac{P_s \times (1 - V_a)}{(\dfrac{P_b}{G_b} + \dfrac{P_s}{G_{se}})} \times (\frac{1}{G_{sb}} - \frac{1}{G_{se}})$$

where V_{ba} = volume of absorbed binder, cm^3/cm^3 of mix
P_b = percent of binder (assumed 0.05),
P_s = percent of aggregate (assumed 0.95),
G_b = specific gravity of binder (assumed 1.02),
V_a = volume of air voids (assumed 0.04 cm^3/cm^3 of mix)

Blend 1: $V_{ba} = \dfrac{0.95 \times (1 - 0.04)}{(\dfrac{0.05}{1.02} + \dfrac{0.95}{2.754})} \times (\dfrac{1}{2.699} - \dfrac{1}{2.754}) = 0.0171$ cm^3/cm^3 of mix

Blend 2: $V_{ba} = \dfrac{0.95 \times (1 - 0.04)}{(\dfrac{0.05}{1.02} + \dfrac{0.95}{2.755})} \times (\dfrac{1}{2.697} - \dfrac{1}{2.755}) = 0.0181$ cm^3/cm^3 of mix

Blend 3: $V_{ba} = \dfrac{0.95 \times (1 - 0.04)}{(\dfrac{0.05}{1.02} + \dfrac{0.95}{2.754})} \times (\dfrac{1}{2.701} - \dfrac{1}{2.754}) = 0.0165$ cm^3/cm^3 of mix

The volume of the effective binder (V_{be}) can be determined from this equation:

$$V_{be} = 0.081 - 0.02931([\ln(S_n)]$$

where S_n = the nominal maximum sieve size of the aggregate blend (in inches)

Blend 1-3: V_{be} = 0.081 - 0.02931([ln(0.75)] = 0.089 cm^3/cm^3 of mix

Finally, the initial trial asphalt binder (P_{bi}) content is calculated from this equation:

$$P_{bi} = \frac{G_b \times (V_{be} + V_{ba})}{(G_b \times (V_{be} + V_{ba})) + W_s} \times 100$$

where P_{bi} = percent (by weight of mix) of binder

W_s = weight of aggregate, grams

$$W_s = \frac{P_s \times (1 - V_a)}{(\dfrac{P_b}{G_b} + \dfrac{P_s}{G_{se}})}$$

Blend 1: $W_s = \dfrac{0.95 \times (1 - 0.04)}{(\dfrac{0.05}{1.02} + \dfrac{0.95}{2.754})} = 2.315$

$$P_{bi} = \frac{1.02 \times (0.089 + 0.0171)}{(1.02 \times (0.089 + 0.0171)) + 2.315} \times 100 = 4.4\% \text{ (by mass of mix)}$$

Blend 2: $W_s = \dfrac{0.95 \times (1 - 0.04)}{(\dfrac{0.05}{1.02} + \dfrac{0.95}{2.755})} = 2.316$

$$P_{bi} = \frac{1.02 \times (0.089 + 0.0181)}{(1.02 \times (0.089 + 0.0181)) + 2.316} \times 100 = 4.4\% \text{ (by mass of mix)}$$

Blend 3: $W_s = \dfrac{0.95 \times (1 - 0.04)}{(\dfrac{0.05}{1.02} + \dfrac{0.95}{2.754})} = 2.315$

$$P_{bi} = \frac{1.02 \times (0.089 + 0.0165)}{(1.02 \times (0.089 + 0.0165)) + 2.315} \times 100 = 4.4\% \text{ (by mass of mix)}$$

Next, a minimum of two specimens for each trial blend is compacted using the SGC. Two samples are also prepared for determination of the mixture's maximum theoretical specific gravity (G_{mm}). An aggregate weight of 4500 grams is usually sufficient for the compacted specimens. An aggregate weight of 2000 grams is usually sufficient for the specimens used to determine maximum theoretical specific gravity (G_{mm}). AASHTO T 209 should be consulted to determine the minimum sample size required for various mixtures.

Specimens are mixed at the appropriate mixing temperature, which is 148°C to 153°C for the selected PG 58-34 binder. The specimens are then short-term aged by placing the loose mix in a flat pan in a forced draft oven at the compaction temperature, 142°C to 146°C, for 2 hours. Finally, the specimens are then removed and either compacted or allowed to cool loose (for G_{mm} determination).

The number of gyrations used for compaction is determined based on the traffic level (Table 6.11).

Table 6.11 Superpave Design Gyratory Compactive Effort			
Design ESALs	Compaction Paramenters		
(millions)	$N_{initial}$	N_{design}	$N_{maximum}$
< 0.3	6	50	75
0.3 to < 3	7	75	115
3 to < 30	8	100	160
≥ 30	9	125	205

The number of gyrations for initial compaction, design compaction, and maximum compaction are:

$$N_{ini} = 8 \text{ gyrations}$$
$$N_{des} = 100 \text{gyrations}$$
$$N_{max} = 160 \text{ gyrations}$$

Each specimen will be compacted to the design number of gyrations, with specimen height data collected during the compaction process. This is tabulated for each Trial Blend. SGC compaction data reduction is accomplished as follows.

During compaction, the height of the specimen is continuously monitored. After compaction is complete, the specimen is extruded from the mold and allowed to cool. Next, the bulk specific gravity (G_{mb}) of the specimen is determined using AASHTO T166. The G_{mm} of each blend is determined using AASHTO T209. G_{mb} is then divided by G_{mm} to determine the %G_{mm} @ N_{des}. The %G_{mm} at any number of gyrations (N_x) is then calculated by multiplying %G_{mm} @ N_{des} by the ratio of the heights at N_{des} and N_x.

The SGC data reduction for the three trial blends is shown in the accompanying tables. The most important points of comparison are %G_{mm} at N_{ini}, N_{des}, and N_{max},. Figures 6.3 through 6.5 illustrate the compaction plots for data generated in these tables. The figures show %G_{mm} versus the logarithm of the number of gyrations.

Densification Data for Trial Blend 1

Gyrations	Specimen 1 Ht, mm	%G_{mm}	Specimen 2 Ht, mm	%G_{mm}	Avg %G_{mm}
5	129.0	85.2	130.3	86.2	85.7
8	127.0	86.5	128.1	87.6	**87.1**
10	125.7	87.3	126.7	88.6	88.0
15	123.5	88.9	124.7	90.1	89.5
20	122.2	89.9	123.4	91.0	90.4
30	120.1	91.4	121.5	92.4	91.9
40	119.0	92.3	120.2	93.4	92.8
50	118.0	93.0	119.3	94.2	93.6
60	117.2	93.7	118.5	94.8	94.3
80	116.0	94.7	117.3	95.8	95.2
100	115.2	95.4	116.4	96.5	**95.9**
G_{mb}	2.445		2.473		
G_{mm}	2.563		2.563		

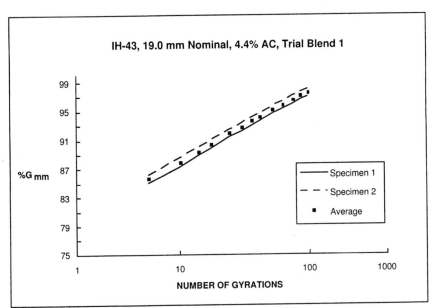

IH-43, 19.0 mm Nominal, 4.4% AC, Trial Blend 1

Figure 6.3 Densification Curves for Trial Blend 1

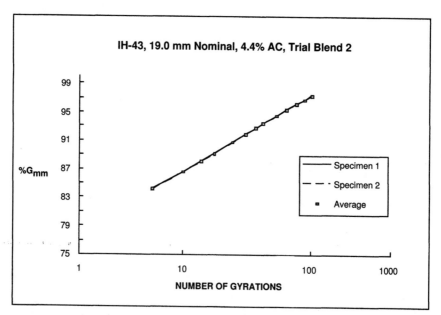

Densification Data for Trial Blend 2

	Specimen 1		Specimen 2		Avg
Gyrations	Ht, mm	%G_{mm}	Ht, mm	%G_{mm}	%G_{mm}
5	131.7	84.2	132.3	84.2	84.2
8	129.5	85.6	130.1	85.6	**85.6**
10	128.0	86.6	128.7	86.6	86.6
15	125.8	88.1	126.5	88.1	88.1
20	124.3	89.2	124.9	89.2	89.2
30	122.2	90.7	122.7	90.8	90.7
40	120.1	91.4	121.5	92.4	91.9
50	119.6	92.7	120.1	92.8	92.7
60	118.7	93.4	119.2	93.5	93.4
80	117.3	94.5	117.8	94.6	94.5
100	116.3	95.3	116.8	95.4	**95.4**
G_{mb}	2.444		2.447		
G_{mm}	2.565		2.565		

IH-43, 19.0 mm Nominal, 4.4% AC, Trial Blend 2

Figure 6.4 Densification Curves for Trial Blend 2

Densification Data for Trial Blend 3

Gyrations	Specimen 1 Ht, mm	%G_{mm}	Specimen 2 Ht, mm	%G_{mm}	Avg %G_{mm}
5	130.9	84.4	129.5	85.2	84.8
8	127.2	85.9	127.3	86.6	**86.3**
10	127.2	86.9	125.9	87.6	87.3
15	125.1	88.3	124.1	89.0	88.7
20	123.7	89.3	122.8	89.9	89.6
30	121.8	90.7	121.0	91.2	91.0
40	120.5	91.7	119.7	92.2	91.9
50	119.6	92.5	118.7	93.0	92.7
60	118.8	93.1	118.1	93.5	93.3
80	117.6	94.0	116.9	94.4	94.2
100	116.7	94.7	116.1	95.1	**94.9**
G_{mb}	2.432		2.442		
G_{mm}	2.568		2.568		

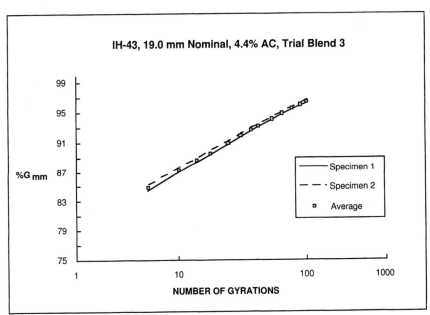

IH-43, 19.0 mm Nominal, 4.4% AC, Trial Blend 3

Figure 6.5 Densification Curves for Trial Blend 3

➤➤ Evaluate Trial Blends

The average %G_{mm} is determined for N_{ini}, (8 gyrations) and N_{des} (100 gyrations) for each trial blend. This data is taken directly from the compaction data tables. The summary of these values for Trial Blends 1, 2, and 3 is shown in Table 6.12:

Table 6.12 Determination of %G_{mm} at N_{ini} and N_{des} for Trial Blends		
Trial Blend	%G_{mm}@N_{ini}	%G_{mm}@N_{des}
1	87.1	95.9
2	85.6	95.4
3	86.3	94.9

The %G_{mm} for N_{max} must also be evaluated. Two additional specimens can be compacted to N_{max} for each of the trial blends or just the selected trial blend can be checked. (In this example, the second approach is utilized. The N_{max} verification, for the example, is discussed later in this chapter.)

The percent of air voids (V_a) and voids in the mineral aggregate (VMA) are determined at N_{des}. The percent air voids is calculated using this equation:

$$\%V_a = 100 - \%G_{mm} @ N_{des}$$

Blend 1:   %Air Voids = 100 - 95.9 = 4.1%
Blend 2:   %Air Voids = 100 - 95.4 = 4.6%
Blend 3:   %Air Voids = 100 - 94.9 = 5.1%

The percent voids in the mineral aggregate is calculated using this equation:

$$\%VMA = 100 - (\frac{\%G_{mm}@ \ N_{des} \times G_{mm} \times P_s}{G_{sb}})$$

Blend 1: $\%VMA = 100 - (\dfrac{95.9 \times 2.563 \times 0.956}{2.699}) = 12.9\%$

Blend 2: $\%VMA = 100 - (\dfrac{95.4 \times 2.565 \times 0.956}{2.697}) = 13.3\%$

Blend 3: $\%VMA = 100 - (\dfrac{94.9 \times 2.568 \times 0.956}{2.701}) = 13.7\%$

Blend	%AC	%G_{mm} @ N=8	%G_{mm} @ N=100	%Air Voids	%VMA
Table 6.13 Compaction Summary of Trial Blends					
1	4.4	87.1	95.9	4.1	12.9
2	4.4	85.6	95.4	4.6	13.3
3	4.4	86.3	94.9	5.1	13.7

Table 6.13 above shows the compaction summary of the trial blends. From this data, an estimated asphalt content to achieve 4% air voids (96% G_{mm} at N_{des}) is determined for each trial blend using this formula:

$$P_{b,estimated} = P_{bi} - (0.4((4-V_a))$$

where $P_{b,estimated}$ = estimated percent binder
P_{bi} = initial (trial) percent binder
V_a = percent air voids at N_{des}
Blend 1: $P_{b,estimated}$ = 4.4 - (0.4(4 - 4.1)) = 4.4%
Blend 2: $P_{b,estimated}$ = 4.4 - (0.4(4 - 4.6)) = 4.6%
Blend 3: $P_{b,estimated}$ = 4.4 - (0.4(4 - 5.1)) = 4.8%

The volumetric (VMA and VFA) and mixture compaction properties are then estimated at this asphalt binder content using the equations below.

For VMA:

$$\%VMA_{estimated} = \%VMA_{initial} + C(4 - V_a)$$

where: $\%VMA_{initial}$ = %VMA from trial asphalt binder content

C = constant (either 0.1 or 0.2)

Note: C = 0.1 if V_a is less than 4.0%

C = 0.2 if V_a is greater than 4.0%

Blend 1: $\%VMA_{estimated}$ = 12.9 + (0.2(4.0 - 4.1)) = 12.9%

Blend 2: $\%VMA_{estimated}$ = 13.3 + (0.2(4.0 - 4.6)) = 13.2%

Blend 3: $\%VMA_{estimated}$ = 13.7 + (0.2(4.0 - 5.1)) = 13.5%

For VFA: $\%VFA_{estimated} = 100 \times \dfrac{(\%VMA_{estimated} - 4.0)}{\%VMA_{estimated}}$

Blend 1: $\%VFA_{estimated} = 100 \times \dfrac{(12.9 - 4.0)}{12.9} = 69.0\%$

Blend 2: $\%VFA_{estimated} = 100 \times \dfrac{(13.2 - 4.0)}{13.2} = 69.7\%$

Blend 3: $\%VFA_{estimated} = 100 \times \dfrac{(13.5 - 4.0)}{13.5} = 70.4\%$

For $\%G_{mm}$ at N_{ini}:

$\%G_{mm\ estimated}$ @ N_{ini} = $\%G_{mm\ trial}$ @ N_{ini} - (4.0 - V_a)

Blend 1: $\%G_{mm\ estimated}$ @ N_{ini} = 87.1 - (4.0 - 4.1) = 87.2%

Blend 2: $\%G_{mm\ estimated}$ @ N_{ini} = 85.6 - (4.0 - 4.6) = 86.2%

Blend 3: $\%G_{mm\ estimated}$ @ N_{ini} = 86.3 - (4.0 - 5.1) = 87.4%

Finally, there is a required range on the dust proportion. This criteria is constant for all levels of traffic. It is calculated as the percent by mass of the material passing the 0.075 mm sieve (by wet sieve analysis) divided by the effective asphalt binder content (expressed as percent by mass of mix). The effective asphalt binder content is calculated using:

$$P_{be} = -(P_s \times G_b) \times (\frac{G_{se} - G_{sb}}{G_{se} \times G_{sb}}) + P_{b,estimated}$$

Blend 1: $P_{be} = -(95.6 \times 1.02) \times (\dfrac{2.754 - 2.699}{2.754 \times 2.699}) + 4.4 = 3.7\%$

Blend 2: $P_{be} = -(95.4 \times 1.02) \times (\dfrac{2.755 - 2.697}{2.755 \times 2.697}) + 4.6 = 3.8\%$

Blend 3: $P_{be} = -(95.2 \times 1.02) \times (\dfrac{2.754 - 2.701}{2.754 \times 2.701}) + 4.8 = 4.1\%$

Dust Proportion is calculated using:

$$DP = \dfrac{P_{0.075}}{P_{be,\,estimated}}$$

Blend 1: $DP = \dfrac{3.1}{3.7} = 0.84$

Blend 2: $DP = \dfrac{2.9}{3.8} = 0.76$

Blend 3: $DP = \dfrac{3.5}{4.1} = 0.85$

The dust proportion criterion (Table 6.14) must typically be between 0.6 and 1.2.

Table 6.14 Dust Proportion of Trial Blends		
Blend	Dust Proportion	Criterion
1	0.84	
2	0.76	0.6 - 1.2
3	0.85	

Tables 6.15 and 6.16 show the estimated volumetric and mixture compaction properties for the trial blends at the asphalt binder content that should result in 4.0% air voids at N_{des}:

Table 6.15 Estimated Mixture Volumetric Properties @ N_{des}						
Blend	Trial %AC	Est. %AC	%Air Voids	%VMA	%VFA	D.P.
1	4.4	4.4	4.0	12.9	69.0	0.84
2	4.4	4.6	4.0	13.3	69.7	0.76
3	4.4	4.8	4.0	13.7	70.4	0.85

Table 6.16 Estimated Mixture Compaction Properties

Blend	Trial %AC	Est. %AC	%G_{mm} @ N=8
1	4.4	4.4	87.2
2	4.4	4.6	86.2
3	4.4	4.8	87.4

Estimated properties are compared against the mixture criteria. For the design traffic and nominal maximum particle size, the volumetric and densification criteria are:

% Air Voids	4.0%
% VMA	13.0% (19.0 mm nominal mixture)
% VFA	65% - 75% ($3 - 30 \times 10^6$ ESALs)
% G_{mm} @ N_{ini}	less than 89%
Dust Proportion	0.6 - 1.2

After establishing all the estimated mixture properties, the designer can evaluate the values for the trial blends and decide if one or more are acceptable, or if further trial blends need to be evaluated.

Blend 1 is unacceptable based on a failure to meet the minimum VMA criteria. Both Blends 2 and 3 are acceptable. The VMA, VFA, D. P., and N_{ini} criteria are met. For this example, Trial Blend 3 is selected as the design aggregate structure.

What could be done at this point if none of the blends were acceptable? Additional combinations of the current aggregates could be tested, or additional materials from different sources could be obtained and included in the trial blend analysis.

Select Design Asphalt Binder Content

Once the design aggregate structure is selected, Trial Blend 3 in this case, specimens are compacted at varying asphalt binder contents. The mixture properties are then evaluated to determine a design asphalt binder content.

A minimum of two specimens are compacted at each of the following asphalt contents:

▲ estimated binder content
▲ estimated binder content ± 0.5%, and
▲ estimated binder content + 1.0%.

For Trial Blend 3, the binder contents for the mix design are 4.3%, 4.8%, 5.3%, and 5.8%. Four asphalt binder contents are a minimum in Superpave mix design.

A minimum of two specimens is also prepared for determination of maximum theoretical specific gravity at the estimated binder content. Specimens are prepared and tested in the same manner as the specimens from the "Select Design Aggregate Structure" section.

The following tables indicate the test results for each trial asphalt binder content. The average densification curves for each trial asphalt binder content are graphed for comparative illustration (Figure 6.6).

Mixture properties are evaluated for the selected blend at the different asphalt binder contents, by using the densification data at N_{ini} (8 gyrations) and N_{des} (100 gyrations). These tables show the response of the mixture's compaction and volumetric properties with varying asphalt binder contents.

The volumetric properties are calculated at the design number of gyrations (N_{des}) for each trial asphalt binder content. From these data points, the designer can generate graphs of air voids, VMA, and VFA versus asphalt binder content (Figure 6.7). The design asphalt binder content is established at 4.0% air voids.

Table 6.17 Densification Data for Trial Blend 3, 4.3% Asphalt Binder

Gyrations	Specimen 1 Ht, mm	%G_{mm}	Specimen 2 Ht, mm	%G_{mm}	Avg %G_{mm}
5	131.3	83.9	131.0	84.7	84.3
8	129.0	85.4	128.8	86.1	**85.7**
10	127.5	86.4	127.4	87.1	86.7
15	125.4	87.8	125.5	88.4	88.1
20	124.0	88.8	124.2	89.3	89.1
30	122.1	90.2	122.4	90.6	90.4
40	120.9	91.1	121.1	91.6	91.4
50	119.9	91.9	120.1	92.4	92.1
60	119.1	92.5	119.4	92.9	92.7
80	117.9	93.4	118.3	93.8	93.6
100	117.0	94.1	117.4	94.5	**94.3**
G_{mb}	2.430		2.440		
G_{mm}	2.582		2.582		

Table 6.18 Densification Data for Trial Blend 3, 4.8% Asphalt Binder

Gyrations	Specimen 1 Ht, mm	%G_{mm}	Specimen 2 Ht, mm	%G_{mm}	Avg %G_{mm}
5	130.4	85.8	130.8	85.5	85.7
8	128.2	87.2	128.8	86.9	**87.1**
10	126.8	88.2	127.4	87.8	88.0
15	124.8	89.6	125.5	89.1	89.4
20	123.5	90.6	124.1	90.1	90.3
30	121.5	92.1	122.1	91.5	91.8
40	120.3	93.0	120.8	92.6	92.8
50	119.3	93.7	119.9	93.3	93.5
60	118.5	94.4	119.0	94.0	94.2
80	117.2	95.4	117.9	94.9	95.1
100	116.4	96.1	117.0	95.6	**95.8**
G_{mb}	2.462		2.449		
G_{mm}	2.562		2.562		

Table 6.19 Densification Data for Trial Blend 3, 5.3% Asphalt Binder

Gyrations	Specimen 1 Ht, mm	%G_{mm}	Specimen 2 Ht, mm	%G_{mm}	Avg %G_{mm}
5	132.0	86.0	132.6	85.8	85.9
8	129.8	87.5	130.4	87.4	**87.4**
10	128.3	88.5	128.9	88.4	88.4
15	126.2	90.0	126.7	89.8	89.9
20	124.8	91.0	125.2	90.9	91.0
30	122.8	92.5	123.2	92.4	92.4
40	121.4	93.5	121.7	93.5	93.5
50	120.3	94.4	120.7	94.3	94.3
60	119.5	95.1	119.0	95.0	95.0
80	118.2	96.1	118.6	96.0	96.0
100	117.4	96.8	117.7	96.7	**96.8**
G_{mb}	2.461		2.458		
G_{mm}	2.542		2.542		

Table 6.20 Densification Data for Trial Blend 3, 5.8% Asphalt Binder

| Gyrations | Specimen 1 | | Specimen 2 | | Avg |
	Ht, mm	%G$_{mm}$	Ht, mm	%G$_{mm}$	%G$_{mm}$
5	130.4	87.4	131.5	87.2	87.3
8	128.6	88.7	129.4	88.6	**88.6**
10	127.4	89.5	128.0	89.6	89.5
15	125.4	90.8	126.2	90.8	90.8
20	124.0	91.9	124.9	91.8	91.8
30	122.4	93.1	123.1	93.1	93.1
40	120.5	94.6	121.3	94.5	94.5
50	119.4	95.5	120.2	95.4	95.4
60	118.9	95.9	119.5	96.0	95.9
80	117.6	96.9	118.2	97.0	96.9
100	116.7	97.7	117.2	97.8	**97.8**
G$_{mb}$	2.464		2.467		
G$_{mm}$	2.523		2.523		

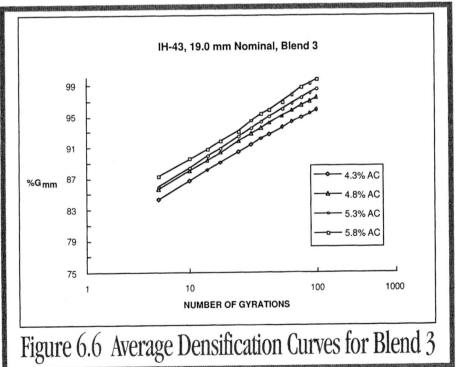

Figure 6.6 Average Densification Curves for Blend 3

Table 6.21 Mix Compaction Properties - Blend 3

%AC	%G_{mm} @ N=8	%G_{mm} @ N=100
4.3%	85.8%	94.3%
4.8%	87.1%	95.8%
5.3%	87.4%	96.8%
5.8%	88.6%	97.8%

Table 6.22 Mix Volumetric Properties at N_{des} - Blend 3

%AC	%Air Voids	%VMA	%VFA	Dust Proportion
4.3%	5.7%	13.7%	58.4%	1.13%
4.8%	4.2%	13.5%	68.9%	0.97%
5.3%	3.2%	13.7%	76.6%	0.85%
5.8%	2.2%	13.9%	84.2%	0.76%

In this example, the design asphalt binder content is 4.9% - the value that corresponds to 4.0% air voids at N_{des} = 100 gyrations. All other mixture properties are checked at the design asphalt binder content to verify that they meet criteria.

Table 6.23 Design Mixture Properties at 4.9% Binder Content

Mix Property	Result	Criteria
Air Voids, %	4.0	4.0
VMA, %	13.5	13.0 min.
VFA, %	71.0	65 -75
Dust Proportion, %	0.9	0.6 - 1.2
%G_{mm} @ N_{ini} = 8	87.2	less than 89

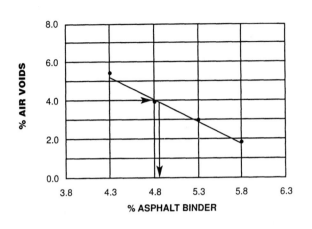

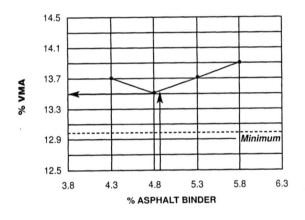

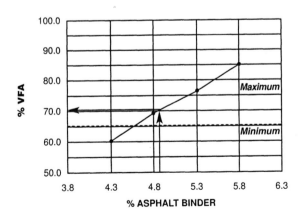

Figure 6.7 Design Mixture Properties at 4.9% Binder Content

➤➤ N_{MAX} Verification

Superpave specifies a maximum density of 98% at N_{max}. Specifying a maximum density at N_{max} prevents design of a mixture that will compact excessively under traffic, become plastic, and produce permanent deformation. Since N_{max} represents a compactive effort that would be equivalent to traffic much greater than the design traffic, excessive compaction will not occur. After selecting the trial blend (#3) and selecting the design asphalt binder content (4.9%), two additional specimens are compacted to N_{max} (160 gyrations).

The table shows the compaction data.

Table 6.24 N_{max} Densification Data for Trial Blend 3, 4.9% Asphalt Binder

	Specimen 1		Specimen 2		Avg
Gyrations	Ht, mm	%G_{mm}	Ht, mm	%G_{mm}	%G_{mm}
5	130.4	85.8	130.8	85.5	85.7
8	128.2	87.2	128.8	86.9	**87.1**
10	126.8	88.2	127.4	87.8	88.0
15	124.8	89.6	125.5	89.1	89.4
20	123.5	90.6	124.1	90.1	90.3
30	121.5	92.1	122.1	91.5	91.8
40	120.3	93.0	120.8	92.6	92.8
50	119.3	93.7	119.9	93.3	93.5
60	118.5	94.4	119.0	94.0	94.2
80	117.2	95.4	117.2	95.4	95.1
100	116.4	96.1	117.0	95.6	**95.8**
125	115.6	96.8	116.2	96.2	96.5
150	115.0	97.3	115.5	96.8	97.0
160	114.5	97.7	115.0	97.2	97.5
G_{mb}	2.495		2.490		
G_{mm}	2.554		2.554		

Blend 3, with %G_{mm} @ N_{max} equal to 97.5, satisfies the Superpave criteria.

Evaluate Moisture Sensitivity The final step in the Superpave mix design process is to evaluate the moisture sensitivity of the design mixture. This step is accomplished by performing AASHTO T 283 testing on the design aggregate blend at the design asphalt binder content. Specimens are compacted to approximately 7% air voids. One subset of three specimens is considered control specimens. The other subset of three specimens is the conditioned subset. The conditioned subset is subjected to partial vacuum saturation followed by an optional freeze cycle, followed by a 24 hour thaw cycle at 60°C. All specimens are tested to determine their indirect tensile strengths. The moisture sensitivity is determined as a ratio of the tensile strengths of the conditioned subset divided by the tensile strengths of the control subset. Table 6.25 shows the moisture sensitivity data for the mixture at the design asphalt binder content. The criterion for tensile strength ratio is 80%, minimum. Trial Blend 3 (82.6%) exceeded the minimum requirement.

The Superpave volumetric mix design is now complete.

Table 6.25 Moisture Sensitivity Data for Blend 3

Sample		1	2	3	4	5	6
Diameter, mm	D	150.0	150.0	150.0	150.0	150.0	150.0
Thickness, mm	t	99.2	99.4	99.4	99.3	99.2	99.3
Dry mass, g	A	3986.2	3981.3	3984.6	3990.6	3987.8	3984.4
SSD mass, g	B	4009.4	4000.6	4008.3	4017.7	4013.9	4008.6
Mass in Water, g	C	2329.3	2321.2	2329.0	2336.0	2331.5	2329.0
Volume, cc (B-C)	E	1680.1	1679.4	1679.3	1681.7	1682.4	1679.6
Bulk Sp Gravity (A/E)	F	2.373	2.371	2.373	2.373	2.370	2.372
Max Sp Gravity	G	2.558	2.558	2.558	2.558	2.558	2.558
% Air Voids(100(G-F)/G)	H	7.2	7.3	7.2	7.2	7.3	7.3
Vol Air Voids (HE/100)	I	121.8	123.0	121.6	121.7	123.4	122.0
Load, N	P				20803	20065	20354
Saturated							
SSD mass, g	B'	4060.9	4058.7	4059.1			
Mass in water, g	C'	2369.4	2373.9	2372.8			
Volume, cc (B'-C')	E'	1691.5	1684.8	1686.3			
Vol Abs Water, cc (B'-A)	J'	74.7	77.4	74.5			
% Saturation (100J'/I)		61.3	62.9	61.3			
% Swell (100(E'-E)/E)		0.7	0.3	0.4			
Conditioned							
Thickness, mm	t"	99.5	99.4	99.4			
SSD mass, g	B"	4070.8	4074.9	4074.8			
Mass in water, g	C"	2373.7	2380.3	2379.0			
Volume, cc (B"-C")	E"	1697.1	1694.6	1695.8			
Vol Abs Water, cc (B"-A)	J"	84.6	93.6	90.2			
% Saturation (100J"/I)		69.5	76.1	74.2			
% Swell (100(E"-E)/E)		1.0	0.9	1.0			
Load, N	P"	16720	16484	17441			
Dry Str. (2000P/(tDp))	Std				889	858	870
Wet Str. (2000P"/(t"Dp))	Stm	713	704	745			
Average Dry Strength (kPa)		872					
Average Wet Strength (kPa)		721					
% TSR		82.6%					

Sieve Size	Nominal Maximum Aggregate Size - Control Point (Percent Passing)									
	37.5 mm		25.0 mm		19.0 mm		12.5 mm		9.5 mm	
	Min.	Max.	Min.	Max.	Min.	Max.	Min.	Max.	Min.	Max.
50.0 mm	100									
37.5 mm	90	100	100							
25.0 mm		90	90	100	100					
19.0 mm				90	90	100	100			
12.5 mm						90	90	100	100	
9.5 mm								90	90	100
4.75 mm									90	90
2.36 mm	15	41	19	45	23	49	28	58	32	67
0.075 mm	0	6	1	7	2	8	2	10	2	10

Minimum and Maximum Boundaries of Sieve Size for Nominal Maximum Aggregate Size
(Minimum and Maximum Percent Passing)

Sieve Size Within Restricted Zone	37.5 mm		25.0 mm		19.0 mm		12.5 mm		9.5 mm	
	Min.	Max.	Min.	Max.	Min.	Max.	Min.	Max.	Min.	Max.
0.300 mm	10.0	10.0	11.4	11.4	13.7	13.7	15.5	15.5	18.7	18.7
0.600 mm	11.7	15.7	13.6	17.6	16.7	20.7	19.1	23.1	23.5	27.5
1.18 mm	15.5	21.5	18.1	24.1	22.3	28.3	25.6	31.6	31.6	37.6
2.36 mm	23.3	27.3	26.8	30.8	34.6	34.6	39.1	39.1	47.2	47.2
4.75 mm	34.7	34.7	39.5	39.5	-	-	-	-	-	-

SELECTION OF MATERIALS

A. Selection of Asphalt Binder
1. Determine project weather conditions using weather database
2. Select reliability
3. Determine design temperatures
4. Verify asphalt binder grade
5. Temperature-viscosity relationship for lab mixing and compaction

B. Selection of Aggregates
1. Consensus properties
   a. Combined gradation
   b. Coarse aggregate angularity
   c. Fine aggregate angularity
   d. Flat and elongated particles
   e. Clay content
2. Agency and Other properties
   a. Specific gravity
   b. Toughness
   c. Soundness
   d. Deleterious materials
   e. Other

C. Selection of Modifiers

II. SELECTION OF DESIGN AGGREGATE STRUCTURE

A. Establish Trial Blends
1. Develop three blends
2. Evaluate combined aggregate properties

B. Compact Trial Blend Specimens
1. Establish trial asphalt binder content
   a. Superpave method
   b. Engineering judgment method
2. Establish trial blend specimen size
3. Determine $N_{initial}$ & N_{design}
4. Batch trial blend specimens

5. Compact specimens and generate densification tables
6. Determine mixture properties (G_{mm} & G_{mb})

C. Evaluate Trial Blends
1. Determine %G_{mm} @ $N_{initial}$ & N_{design}
2. Determine %Air Voids and %VMA
3. Estimate asphalt binder content to achieve 4% air voids
4. Estimate mix properties @ estimated asphalt binder content
5. Determine dust-asphalt ratio
6. Compare mixture properties to criteria

D. Select Most Promising Design Aggregate Structure for Further Analysis

III. SELECTION OF DESIGN ASPHALT BINDER CONTENT

A. Compact Design Aggregate Structure Specimens at Multiple Binder Contents
1. Batch design aggregate structure specimens
2. Compact specimens and generate densification tables

B. Determine Mixture Properties versus Asphalt Binder Content
1. Determine %G_{mm} @ $N_{initial}$ & N_{design} & $N_{maximum}$
2. Determine volumetric properties
3. Determine dust-asphalt ratio
4. Graph mixture properties versus asphalt binder content

C. Select Design Asphalt Binder Content
1. Determine asphalt binder content at 4% air voids
2. Determine mixture properties at selected asphalt binder content
3. Compare mixture properties to criteria

IV. EVALUATION OF MOISTURE SENSITIVITY OF DESIGN ASPHALT MIXTURE USING AASHTO T283

MEMBERS OF THE ASPHALT INSTITUTE
(As of August 2001)

The Asphalt Institute is an international, nonprofit association sponsored by members of the petroleum asphalt industry to serve both users and producers of asphalt materials through programs of engineering service, research and education. Membership is available to refiners of asphalt from crude petroleum; to processors manufacturing finished paving asphalts and/or non-paving asphalts but not starting from crude petroleum; and to companies working specifically with asphalt related raw material or asphalt additives.

*Akzo Nobel Surface Chemistry LLC, Willowbrook, Illinois
All States Asphalt, Inc., Sunderland, Massachusetts
Alon USA, Dallas, Texas
*Andrie Inc., Muskegon, Michigan
*Arr-Maz Products, Winter Haven, Florida
Ashwarren International Inc., Mississauga, Ontario, Canada
Asphalt Materials, Inc., Indianapolis, Indiana
Asphalt Processors Incorporated, Barbados, West Indies
Associated Asphalt, Inc., Roanoke, Virginia
ATOFINA Petrochemicals, Inc., Houston, Texas
*ATOFINA Petrochemicals, Inc., Houston, Texas
*BASF Corporation, Charlotte, North Carolina
Bitumar Inc., Montreal, Quebec, Canada
Bituminous Products Company, Maumee, Ohio
*Bouchard Coastwise Management Corp., Hicksville, New York
BP, Whiting, Indiana
Caltex Corporation, Sydney, Australia
Canadian Asphalt Industries Inc., Markham, Ontario, Canada
Centennial Gas Liquids L.L.C., Denver, Colorado
Chevron Products Company, San Ramon, California
CITGO Asphalt Refining Company, Plymouth Meeting, Pennsylvania
*Coastal Towing, Inc., Houston, TX
Colas, S.A., Paris, France
Consolidated Oil & Transportation Co., Inc., Englewood, Colorado
*Dexco Polymers (A Dow/ExxonMobil Partnership), Houston, Texas
*DuPont, Wilmington, Delaware
*Dynasol L.L.C., Houston, Texas
EMCO Limited, Building Products, LaSalle, Quebec, Canada
*EniChem Americas, Inc., Houston, Texas
Equiva Trading Company, Burbank, California
Ergon Asphalt & Emulsions, Inc., Jackson, Mississippi
ExxonMobil Lubricants & Petroleum Specialties Company, Fairfax, Virginia
ExxonMobil Lubricants & Petroleum Specialties Company (International), Fairfax, Virginia
Frontier Terminal & Trading Company, Tulsa, Oklahoma
GAF Materials Corporation, Wayne, New Jersey
Golden Bear Oil Specialties, Los Angeles, California
Gorman Asphalt, LTD, Rensselaer, New York
Gulf States Asphalt Co., L.P., South Houston, Texas
*Heatec, Inc., Chattanooga, Tennessee
Hunt Refining Company, Tuscaloosa, Alabama
Husky Oil Marketing Company, Calgary, Alberta, Canada
IKO, Chicago, Illinois
Imperial Oil, Toronto, Ontario, Canada
Isfalt A.S., Uskudar, Istanbul, Turkey
Jebro Inc., Sioux City, Iowa
Koch Materials Company, Wichita, Kansas
*KRATON Polymers, Houston, Texas
Marathon Ashland Petroleum LLC, Findlay, Ohio
E. A. Mariani Asphalt Co., Inc., Tampa, Florida
Mathy Construction Company, Onalaska, Wisconsin
McAsphalt Industries Ltd., Scarborough, Ontario, Canada
Moose Jaw Asphalt Inc., Moose Jaw, Saskatchewan, Canada
Murphy Oil USA, Inc., Superior, Wisconsin
Nynäs Bitumen, Brussels, Belgium
Oldcastle Materials Group, Washington, D.C.
Paramount Petroleum Corporation, Paramount, California

*Penn Maritime, Inc., Stamford, Connecticut
Petro-Canada Inc., Oakville, Ontario, Canada
*Petro-Nav Inc., Montreal, Quebec, Canada
Petroleo Brasileiro, S.A. - Petrobras, Rio de Janeiro, Brazil
Pioneer Oil LLC, Parker, Colorado
RECOPE, Cartago, Costa Rica
Repsol Productos Asfalticos, S.A., Madrid, Spain
*Rohm and Haas Company, North Andover, Massachusetts
*Safety-Kleen, Oil Recovery Division, Elgin, Illinois
San Joaquin Refining Co., Inc., Bakersfield, California
Sargeant Marine, Inc., Boca Raton, Florida
Seneca Petroleum Co., Inc., Crestwood, Illinois
Shell Canada Products, Montreal, Quebec, Canada
Shell International Petroleum Company Limited, London, England
SK Corporation, Seoul, Korea
Southland Oil Company, Jackson, Mississippi
Suit-Kote Corporation, Cortland, New York
Terry Industries, Inc., Hamilton, Ohio
Tesoro Petroleum Corporation, Anacortes, WA
*Texaco Refining and Marketing, Inc., Marrero, Louisiana
Tosco Refining Company, Phoenix, AZ
Trumbull Products (Division of Owens Corning), Toledo, Ohio
Ultramar Diamond Shamrock, San Antonio, Texas
*Ultrapave Corporation, Resaca, Georgia
United Refining Company, Warren, Pennsylvania
U.S. Oil & Refining Company, Tacoma, Washington
Valero Energy Corporation, San Antonio, Texas
Warden Modified Asphalt, Harrisburg, Pennsylvania
YPF S.A., Buenos Aires, Argentina

* Affiliate Member

LaVergne, TN USA
17 February 2011
216941LV00005B/2/P